Witt · Reed · Peakall   A Spider's Web

Peter N. Witt
Charles F. Reed · David B. Peakall

A Spider's Web

Problems in Regulatory Biology

With 47 Figures

Springer-Verlag New York Inc. 1968

PETER N. WITT

Division of Research, North Carolina Department of Mental Health · Raleigh, N.C.

CHARLES F. REED

Department of Psychology, Temple University · Philadelphia, Pa.

DAVID B. PEAKALL

Division of Biological Sciences, Langmuir Laboratories, Cornell University · Ithaca, N.Y.

Title of the Original German Edition:

PETER N. WITT: Die Wirkung von Substanzen auf den Netzbau der Spinne als biologischer Test
Springer-Verlag Berlin · Göttingen · Heidelberg 1956

Preface

"Gradually, a faint brightness appeared in the east, and the air, which had been very warm through the night, felt cool and chilly. Though there was no daylight yet, the darkness was diminished, and the stars looked pale. The prison, which had been a mere black mass with little shape or form, put on its usual aspect; and ever and anon a solitary watchman could be seen upon its roof, stopping to look down upon the preparations in the street ... By and by the feeble light grew stronger, and the houses with their sign-boards and inscriptions stood plainly out, in the dull grey morning ... And now, the sun's first beams came glancing into the street; and the night's work, which, in its various stages and in the varied fancies of the lookers-on had taken a hundred shapes, wore its own proper form — a scaffold and a gibbet ..."
(The Complete Works of Charles Dickens, Harper & Brothers, New York and London, Barnaby Rudge, Vol. II, Chapter XIX, page 164.)

Dickens describes an activity which takes place in the early morning hours, just before sunrise. As the day begins and people start to go about their business and get ready to watch the hanging, the hangman is ready with the gallows. — This same early hour of the day sees the spider weaving the last strands of its orb-web and settling down in the middle to wait for flying insects, which it hopes to catch in the trap. — Both apparently select for their activity that time at which ordinary individuals like to sleep or awake drowsily. Both apparently prefer to be unobserved during the process, but ready and waiting when others come out. — The odd hours at which webs are constructed led to the first drug experiments, which in turn led to more investigation of web-construction. The course which investigations followed after the first impulse, and the thoughts which observation of the web building stimulated, as well as some general considerations on regulation and interaction of bodily and "mental" processes are the subject of this book.

Contents

I. The Story of the Drug Web

In the year 1948 Dr. HANS M. PETERS, professor of zoology, asked advice from the Pharmacology Department at the University of Tuebingen. He had tried to take movie pictures of spiders during web construction; this usually takes place around 4 o'clock in the morning, a bad hour for the movie crew to work. It was hoped that with the help of stimulant drugs, web building time could be shifted forward. At that time I investigated drugs which attacked primarily central nervous system functions. The use of dextro-amphetamine, strychnine, and morphine was recommended as a first approach to changing the behavior of an animal which had not yet been tried in its reaction to drugs. Dr. PETERS soon reported that the experiments had resulted in badly distorted webs, built at the usual time. A movie was taken of a spider building a web under the influence of amphetamine and resulting in a most irregular pattern.

At this point the experiments lost interest for the zoologists, but gained interest for psychopharmacology. A way of web photography and some measures for the geometry of the web had already been developed by PETERS. It was found that there was no difficulty in giving spiders drugs by mouth if the right additive like sugar water was used. As methods to test effects of drugs on behavior were poorly developed and rarely quantitative at that time, it became of interest to work out a method using the web as a measurable record of behavior, and establishing its changes as the consequence of drug application. It became soon apparent that a measurable web of some kind was built even by severely disturbed animals.

The first drugs tried were caffeine, amphetamine, scopolamine, strychnine, most of which showed statistically significant changes in distinct web proportions. A standard method of web photography and measurement was established and applied to compare drug and control webs statistically.

Though it seemed that the web test as a method discriminated reasonably well between different drugs, as a research tool it remained unsatisfactory: it could only do screening, which means distinguish between different groups of drugs, but offered no explanation of the mechanism of drug action. But even large scale screening was made difficult by the enormous number of measurements and calculations which had to be performed. For example, one web of an adult female *Araneus diadematus* Cl. may easily contain 35 radii and 40 spiral turns. The catching part of the web, its main component, possesses 1400 intersections which could be taken as measuring points. 20 webs before and 20 webs after drug application have to be measured for statistical comparison; this means measurement of 56,000 points in two coordinates. From that the regularity, size and shape measures have to be generated, a procedure which demands extensive manipulation of all figures. To save effort, only those measures which seemed of special interest for a particular drug were taken and processed, which made results of different drug experiments hardly comparable.

Still there was one application: it was more and more discussed that the acute hallucinatory stages of mental diseases could be attributed to a foreign substance in the possibly abnormal metabolism in the patient's body. Substances like LSD 25 with its effectiveness in unbelievably small amounts, 40 μg per person, or mescaline, adrenochrome, and others which produce hallucinatory excitatory states in normal

Fig. 1. Web of an adult female *Araneus diadematus* Cl. This web was built in the laboratory in a 50 × 50 cm aluminium frame in the early morning hours. Later the spider was removed, the thread covered with a thin layer of white spray paint, and a perpendicular measure added. The photograph was taken in the opening of a black box, with fluorescent lighting from the sides, on high contrast film

people, were candidates for such a substance. Experiments were carried out at that time to find out whether in the body fluids (serum, cerebrospinal fluid, urine) of acutely hallucinating patients such "abnormal" substances could be discovered with the spider test. Results were contradictory, and no clear presence of a foreign substance in certain types of mental patients could be established. Further experiments at that time were abandoned, not least for the reason that in order to make a meaningful statement, many different fractions of body fluids at different doses had to be tested and this involved a prohibitive amount of work.

In the following years a number of experiments were undertaken to further clarify changes in drug webs. For example, it was not clear whether there was a difference between experimenting with young and old spiders. Different species could react in different ways. Many spiders had lost a leg in the experiments. Does a smaller number of legs, 6 or 7, affect web geometry? A great number of experiments were undertaken to clarify all such questions, and they are discussed in a later chapter.

It also became clear that the amount of silk used in the web may be a limiting factor in web construction. A starved spider might produce less silk, or the same amount of silk, a thicker or thinner thread to be spread out in a smaller web, or a larger, wider meshed web; a great number of regulations could possibly take place in the situation of food deprivation, a tendency to save silk being counteracted by the necessity to trap food. Experiments indicated that the web pattern was severely influenced by the silk supply. However, it was never clear whether the spider would under normal circumstances empty all its glands, and where different threads came from. At this point Dr. DAVID PEAKALL joined the spider explorations and with his background of physical chemistry and protein chemistry, started to investigate silk production at the glandular level. He established in a number of experiments the location of a regulatory mechanism for silk: acetylcholine-cholinesterase on the gland surface regulated the speed of silk synthesis through a possibly neural mechanism; and in addition, speed of silk production was dependent on the filling stage of the gland. Dr. PEAKALL had to repeat much of the ground work on gland function and silk origin too. Reports in the literature concerned with the characteristics of silk of one spider were done without taking into consideration the different composition of the threads which, for example, contribute to one web. It was Dr. PEAKALL who seemed therefore the person to discuss the silk and glands in this book on the basis of his own findings. Among others, the results of his work permit to study interaction of a behavioral variable (web-building) with a biochemical mechanism (silk synthesis) and maybe establish the spider as a useful model for such studies.

As mentioned before, one of the limiting factors in drug as well as other research with spider webs was the abundance of data to be processed. A great number of webs have to be measured to get statistical results of reasonable reliability, telling us about a change attributed to a specific drug. Each drug had to be investigated at different dose levels and different times. Dr. CHARLES REED with a background in psychology and specific interest in the mathematical-statistical approach to biological problems became interested in spider webs at this point. He suggested using the recently introduced computer methods for the evaluation of the web pattern. With the help of Mr. JONES from the IBM Corporation, Dr. REED and I developed a computer program to evaluate spider web measurements. This permitted quick assessment of many data. Through analysis of intercorrelations of web parameters, a reduction of measurements to a minimum number became possible. For the last 4 years Dr. REED and I have closely cooperated in web-research, and his thinking can be found in two chapters.

Though the method of harvesting ready built webs in the morning, photographing the pattern, and measuring afterwards was most convenient, it became apparent that one could gain additional information by asking questions of spiders in action. The early morning hours make this difficult. Also the thread which is nearly invisible as it extrudes from the spinnerets of the spider, and the very rapid movements of the animal, make observations frustrating. An attempt was made to take movie pictures of a spider during web construction; hopefully such pictures show thread as it emerges from the spinnerets and is positioned into the pattern. First movie pictures have been obtained in 1966 with the help of Mr. TERRY BARNUM, and much more of this and analysis of the movies with slowed-down projection is under way. Another approach to analysis of the building process is series photography with a camera which takes pictures at regular intervals during web construction. Mr. SAMUEL BAYS and others have started to apply such a procedure successfully in 1966, and mathematical-statistical analysis is under way to determine the probability with which a thread is put in a certain position (see page 82). Correct prediction of a number of threads can be used for analysis of the cues which a spider might use during construction.

The spider web, its pattern, construction and use, have been employed in many more ways to investigate problems of invertebrate behavior and behavior in general. The relationship of the function of certain areas of central nervous system to changes in web building behavior has been under investigation in this laboratory for several years. Minute lesions (0.02 mm diameter) in the central nervous system are produced by means of a laser beam, and comparison of an individual's web-pattern before and after such treatment aids in the analysis of central nervous system function. Another approach to the study of spider behavior and its change through experience has recently been elaborated by Dr. LOUIS LE GUELTE in Raleigh, N.C. and Nancy, France: turning around the web of *Zygiella* makes it difficult for the animal to find the way from a prey in the web back to its retreat. In repeated experiments the animal improves the speed with which it finds the way, and such change is dependent on the number of trials as well as their spacing in time. Such experimentation may open a new avenue for quantitative assessment of the establishment of a memory trace and its elimination, with and without drug interference.

If we follow the urge to satisfy the curiosity of the mind about the problems of nature which surround us every day of our life, there are many ways in which one can proceed. In this book we describe the way in which a small group of people used one particular animal, the spider, in one of its special functions, web-building, as the way. In the nearly 20 years of our special approach, the work was never done in isolation: much thought and experimentation has preceded our work, and we have gratefully profited from it by using newly invented techniques to reaffirm and elaborate older findings. Investigators from all over the world have joined us, in spirit or in person, to travel for some time on the same road, and much thanks are due to their stimulating cooperation. The acquaintance of other workers in the field has only been made through their writings, and their influence is most gratefully acknowledged. Rather than singling out an always incomplete number of individuals who exchanged thoughts with us and influenced us, we mention none and leave it to the text and reference list to inform the reader.

This book is selective. It does not claim completeness on spider anatomy or even web-building. Good comprehensive books exist which quote every description, observation, and experiment on spiders which has been printed. We have selected to describe our personal approach, our hypotheses, and our own laboratory evidence as an example of how it can be done. And while this goes into print, new evidence appears and new thoughts turn up. Like the previous book in 1956, this constitutes a progress report rather than a finished product. Its value and general interest lie in the fact that an animal model and methods for its exploration have been developed by us which appear valuable for many explorations of problems in regulatory biology.

II. The Silk Glands

Introduction

The silk glands of the spider are attractive models for study since they are discrete organs which have the sole function of the rapid production of single proteins. Evolution in the *Arachnidae* has been such as to make the greatest use of silk fibroins. The spider catches its food in a silken web, swathes its prey in a silken bag, and wraps its eggs in a silken cocoon. Wherever the spider goes, it lays a silken thread and throughout its life its contact with the outside world is via its silk.

Araneus has at least five sets of silk glands. Each is modified in the structure of the gland and in the chemical composition and physical properties of the fibroin for a specific purpose. Every event in the life of the spider, whether it is studied at the behavioral or metabolic level, involves the use of these glands. The advantage at the behavioral level is that the web makes a permanent record which is left for detailed analysis. At the biochemical levels the glands are of interest because they synthesize a single structural protein at a rapid rate.

The anatomy of the spider will not be considered in any detail except for two areas which are of special interest to us. These are the nervous system (see Chapter 5) and the silk glands. For detailed treatment of other anatomical aspects the reader is referred to SAVORY (1928), COMSTOCK (1948), and GERTSCH (1949). However, since the basic anatomy of the spider may not be familiar to all readers, it will be briefly described.

A simplified drawing of the anatomy of *Araneus*, made from a series of micro-photographs, is shown in Fig. 2. Spiders, unlike insects, have only two major body compartments, the cephalothorax and the abdomen. The cephalothorax contains the central nervous system, the eyes, legs, and poison glands. The abdomen contains the heart, lungs, and silk glands. The digestive system is found in both segments. Most of the digestion of the spider's food takes place not inside the spider but inside its victim. The spider injects digestive enzymes into its prey; after the contents of the body have become liquid, they are sucked up. This is accomplished by means of a sucking stomach: not a stomach in the true sense of the word but rather a pump. Recent analysis of movie pictures by WITT and BARNUM shows that a rather large ball of silk disappears — presumably into the spider's mouth — in $1/8$ to $1/2$ second. If confirmed, this would be interpreted as a possibility that silk is swallowed before it is completely digested. The pre-digested material then passes, via the narrow pedicel, into the abdomen, where most of the absorption from the intestine takes place. The branched tubes of the intestine occupy almost all the upper and hind parts of the abdomen. This extensive area is shown in yellow in the diagram. It acts not merely as a digestive organ but also as food reservoir.

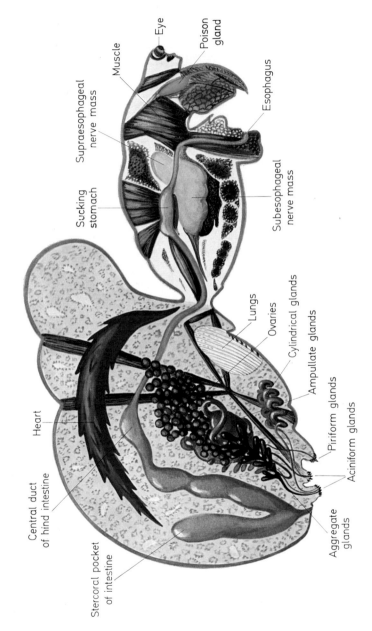

Fig. 2. Simplified anatomy of the spider drawn by WAYNE TRIMM from microphotographs. (Reproduced by permission from the Conservationist, Vol. 20, p. 19 (1966))

The musculature of the abdomen is relatively limited. There are the longitudinal ventral muscles that control the movement of the spinnerets and a framework of muscles that are attached to the body wall. These muscles may be of importance in the movement of silk within the silk gland (see p. 15). There is but one circuit of blood and this is not closed. The heart (shown in red in the diagram) is simple, without division into chambers, the arteries are few and there are no capillaries to the various organs. In fact, the organs of the body are suspended more or less freely in a bath of

Table 1. *Number and position of each type of silk gland, and function of the silk. The number of glands is based largely on the work of* WARBURTON *(1890); the modification below the line on the studies of* SEKIGUCHI *(1952). Function of the silk is based on* APSTEIN *(1889),* WARBURTON *(1890),* COMSTOCK *(1948),* SEKIGUCHI *(1952),* KASTON *(1964), and* PEAKALL *(1964b)*

Type of silk gland	Classification of spinnerets			Function of silk
	Anterior	Median	Posterior	
Ampullate	2	2	—	Drag-line, scaffolding of webs, base-line of catching spiral(?)
Aggregate	—	—	6	Viscid thread of catching spiral
Cylindrical or tubuliform	—	2	4	Egg sac
Piriform	200	—	—	Attachment discs, thin components of frame thread and radii?
Aciniform	—	200	200	Swathing bands
Aggregate	—	—	4	Viscid thread of catching spiral
Flagelliform[a]	—	—	2	Base-line of catching spiral

[a] Named coronata independently by PETERS (1955).

blood. This has two points of consequence to the experiments described herein, the first is that it allows for the ready and undamaged removal of silk glands from the spider for *in vitro* experiments and the second is that materials injected into the abdomen are rapidly available to the silk glands.

The spiders *Araneus diadematus* and *A. sericatus* have five or six sets of silk glands. These are listed in Table 1 and will now be considered separately.

A. Structure and Functions of Silk Glands
The Ampullate Glands

There are two pairs of these glands. The larger pair is located chiefly in the rostral part of the abdomen and is connected to the anterior spinnerets. The second, smaller, pair of ampullate glands (shown in Fig. 3), is situated on either side of the ovaries and is connected to the medial spinnerets. An ampullate gland can be conveniently divided into three parts. A long, thin, looped duct runs from the spinnerets to the wide sac-like region which, in turn, narrows to a long thin tail coiled around the sac. The ampullate gland is usually depicted as having a wide epithelium in the central sac of the gland and narrow in the coiled tail; in *Araneus diadematus* and *A. sericatus* this is not correct: the epithelia of both parts are approximately equal in thickness (see Table 2). The fine structure of the tail of the gland is shown in Fig. 5. As would be expected in view of the rapid rate of protein synthesis of which this gland is capable, there is a large amount of ergastoplasm. A considerable number of droplets of pre-formed protein can be seen in the cell. No Golgi apparatus has been found in the cell, and the protein droplets are apparently secreted directly into the lumen. The

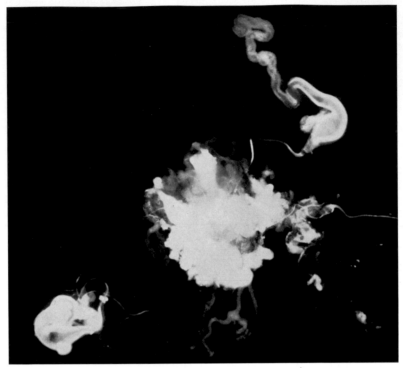

Fig. 3. Two ampullate glands of *Araneus sericatus* attached to spinnerets. Gland on upper right is uncoiled to show proportions of the three parts; gland on left shows original configuration. Magnification ×6

ampullate gland differs from other secretory glands that have been studied, i.e. the pancreas (PALADE et al., 1961) and the silk gland of the silkworm (AKAI and KOBAY-ASHI, 1965), where the protein is concentrated by the Golgi apparatus. The changes

Table 2. *Dimensions of the duct, sac and tail of an ampullate gland (female Araneus sericatus, 60 mg)*

	Stretched length of section (mm)	Thickness of epithelium (μ)	Diameter of lumen (μ)	Calculated volume of	
				epithelium (μl)	lumen (μl)
Duct	10	9.2 ± 0.5	48.3 ± 2.1	0.1	0.2
Sac	1.5	39.1 ± 6.5	800 (max.)	3.2	15.1
Tail	7	35.0 ± 5.3	170 ± 29	1.6	1.6

in the fine structure of the ampullate gland following secretion of the pre-formed protein are described later.

The duct is thin-walled. The wall of the duct appears structureless under the light microscope. Normal cytoplasm and nuclei of cells, considered to be part of the duct, are seen in contact with the duct wall. Under the electron microscope, the ring is seen to consist of parallel intra-cellular canaliculi with many small side tubes which appear to occupy most of the space between the canaliculi. These canaliculi do not open into the lumen of the duct but are separated from it by a distinct, continuous membrane. At the basal side of the ring, the canaliculi terminate at definite, individual membranes which show more clearly than do the membranes of the luminal portion of the canaliculi. There are no signs of normal cytoplasmic organelles — mitochrondira, ribosomes — in this portion of the cell (BELL and PEAKALL, in press). It is possible that the function of this section of the gland is to remove water as well as to deliver the protein to the spinnerets. There is some evidence that removal of water plays a role in the transformation of the liquid protein into a thread. Liquid silk removed from the dissected gland loses 30—40% of its weight when heated for 15 minutes at 70°. Reeled silk, on the other hand, loses little weight under these conditions, although it is possible that water is lost rapidly from the fine fiber as it leaves the spigot. Attempts were made to minimize this by reeling under conditions of high humidity and it can be readily shown that the fiber can be reeled from the spider under water. The idea that water is removed in the duct is without direct experimental proof but could be correlated with the lack of a Golgi apparatus in the secretory cycle.

Protein synthesis occurs largely, but not exclusively, in the tail of the gland; this has been demonstrated by means of autoradiography. Spiders were injected with labeled alanine and an hour later the glands were dissected out and fixed. The counts of the incorporated alanine per unit area were 9.1 ± 2.3 for the epithelium of the sac and 43.1 ± 6.1 for the epithelium of the tail. Both figures represent means of fifty counts with standard deviations. These counts give an index of the amount of new protein synthesis occurring over a one hour period. The method of autoradiography is discussed later. This difference in the rate of synthesis in different parts of the gland has also been described in the silk gland of the silkworm (LUCAS et al., 1958).

The Aggregate Glands

The aggregate gland is an irregularly lobulated structure found in close proximity to the ampullate gland. The duct is short compared to that of the ampullate gland and there is no apparent division of the gland into sac and tail portions. Autoradiographic studies show no difference in the amount of incorporation in various parts of the gland. The average of 20.9 ± 5.1 counts per unit area was obtained under conditions comparable with the values quoted for the ampullate gland. I have no histological evidence in *Araneus diadematus* and *sericatus* to support the sub-division of the aggregate glands into aggregate and flagelliform glands as has been proposed by SEKIGUCHI (1952) and PETERS (1955).

Function of Silk Glands Used in Web-Building

Specific purposes for the silk from each set of glands have been proposed and these are listed in Table 1. Inferences are based upon observation of the spinneret being used in a specific task and on a knowledge of the gross anatomy of the glands. We have confirmed visual observation by means of cine-photography. Also, by serial sectioning spiders after a specific silk has been used, I have been able to correlate type of gland with type of fibroin. Sections have been examined from spiders at rest, after the completion of the dry scaffolding of the web, after completion of the entire web, and after thread has been artifically pulled from the spider. Microphotographs of these sections have been published previously (PEAKALL, 1964b). It was found that both the construction of the scaffolding of the web and the pulling of thread from the spider depleted the silk from the ampullate glands. The aggregate glands were found to be full at the completion of the dry scaffolding of the web (i.e. at the completion of the provisional spiral) and depleted when the web was completed. Thus, the involvement of the aggregate glands in the formation of the catching spiral is clear. The use of these glands in the production of the viscid thread had been proposed as early as 1890 by WARBURTON who based his conclusions on the fact that aggregate glands are found only in spiders that build orb-webs. However, no supporting evidence had previously been obtained.

The fact that the catching spiral consists of more than one component can be clearly seen by examination under a low-powered microscope. The thread consists of relatively evenly-spaced droplets on a base thread (see for example SAVORY, 1952, plate 5, p. 37). RICHTER (1956) presented photographic evidence for the existence of three layers, a structural thread, a viscid layer, and an outer mucuous layer. However, the value of this evidence cannot be assessed since neither the experimental technique nor the name of the spider used was given. KASTON (1964), states that when the outer layer is washed off, the fibroin quickly becomes brittle; but again no experimental details are given.

Still left open is the question of the source of silk for the central supporting thread of the sticky spiral. SEKIGUCHI (1952) and PETERS (1955) have suggested that the structural thread comes from the flagelliform glands. KOENIG (1951) allowed spiders to build webs to the stage where all but the last three radii had been laid down. Then he continously eliminated the last three radii. He found that the spider continued to lay down radii until it had built two and one-half times the normal number of radii. Clearly, then, some silk remains in the ampullate glands at the end of normal radii building. This conclusion is in agreement with our histological data. KOENIG then found that the sticky spiral, which the spider finally built after repeated elimination of radii by the experimenter, was very small. It could be inferred that this small size is due to the shortage of material in the ampullate glands. Unless one assumes that the number of radii is crucial for the size of the spiral, this is not conclusive evidence (see further discussion on page 75).

Occasionally both the piriform and aciniform glands, are used in conjunction with the ampullate gland. WARBURTON (1890) describes that "upon rare occasions, the whole battery of tubes seems to be brought into play, the posterior spinnerets contributing their quota to the strengthening of the line". Thus, the "trailing line" may on occasion be strengthened by contributions from the piriform and aciniform. However, examination of motion pictures of web-building made by WITT and BARNUM do not suggest that threads from either of these sets of small glands are regularly used as a source of the fine thread of the scaffolding as suggested by KASTON (1964).

SEKIGUCHI (1955a) has studied the differences in the spinning glands between adult male and adult female spiders for several species, including in the genus we have studied, *Araneus ventricosus*. SEKIGUCHI found no differences between the three genera — *Araneus*, *Argiope*, and *Tetragnatha*. He reported that the aggregate glands become vestigial in the adult male and that a male in the laboratory did not spin a web after its last molt. SEKIGUCHI's observation that an adult male is seldom found sitting in an orb-web is confirmed by the author's experience during collection of *Araneus sericatus*. The feeding habits of the male at this stage in the wild do not appear to have been studied. In view of the fact that it may be eaten by the female after copulation, not many specimens are available for study.

Structure and Function of the Other Silk Glands

The aciniform glands in *Araneus diadematus* and *A. sericatus* are not similar to multiple sacs emptying into a common duct as described by COMSTOCK (1948) for the family *Aranea*. Rather, each spinning tube has a single small gland which lies just inside the wall of the spinneret. A microphotograph of this gland has been shown elsewhere (PEAKALL, 1964b). The piriform glands are very similar in appearance, but are found only on the anterior spinneret. The functions of these two sets of glands have been separated by serial sectioning of the spider after it had swathed several flies. Since the piriform glands were empty and the aciniform glands full, the piriform glands presumably contribute to the swathing material.

The cylindrical glands provide the silk for the egg cocoon. They have the appearance of coiled cylinders and the diameter of these glands is comparatively uniform throughout their length. In the young female the structure of the epithelial cells is normal, with a distinct nucleus and cytoplasm. The lumen is filled with protein. In the adult female the epithelial cells are structureless under the light microscope. Treatment of the sections with RNase followed by staining with Azure B shows, by the amount of dye absorbed, the presence of considerable quantities of DNA. This was confirmed by the low uptake of dye following treatment with DNase. However, there is no localization of the DNA within the gland. The epithelium also contains considerable amount of protein, but again there is no visible localization. In the adult female the structureless cylindrical glands may be seen either full or empty. Examination of spiders at periods up to 30 days after cocoon formation shows that the glands do not refill (PEAKALL, 1965b). It appears that the silk for the cocoon is made early in the life-cycle of the spider after which atrophy of the gland occurs. The regulation of this gland is thus completely different from that of the other silk glands (see also page 25). SEKIGUCHI (1955a) found that the cylindrical glands are absent in the adult male; this is not surprising in view of the function of this gland.

Chemical and Physical Properties of Silk

All spider silks are proteins, but the chemical and physical properties of the different silks from the various silk glands of the same spider vary considerably. The evolution of silks in spiders has been such as to make the maximum use of this material.

Amino acid composition. LUCAS et al. (1958, 1960) have summarized the values for amino acid composition of various fibroins. These determinations cover three orders (Neuroptera, Lepidoptera, and Hymenoptera) in the class Insecta and several species of the class Arachnida.

In all these fibroins, the amino acids with small side-chains (glycine, alanine, and serine) are dominant, although the variation of both the total of these residues and the ratio of them, is considerable. The lowest total of alanine, glycine, and serine is 42.6% *(Arctia caja)* and the highest 94.6% *(Anaphe moloneyi)*. The fibers of Antheraea contain 42% alanine and those of Chrysopa 41% serine. The structural requirement of these fibers appears to necessitate a high proportion of these amino acids since they are found in fibroins over such a wide taxonomic range. LUCAS et al. (1960) have found no precise relationship between the amino acid composition of the various fibers and their biological classification. WARWICKER (1960) has found some relationship between content of short side-chain amino acids (alanine, glycine, and serine) and the X-ray diffraction pattern of the fibroin. SMITH (1966) has calculated the amino acid composition of the 'average' protein. On this basis he finds that silk (from *Bombyx mori*) is one of the most unusual of proteins. Plotting the curve of the variation from the average he found silk three standard deviations away from the mean.

Taxonomic considerations of silk proteins. The term 'protein taxonomy' has been coined to describe the study of the sequence of amino acids comprising the proteins of an organism and the comparison of them between species. CRICK (1958) has argued that these sequences are the most delicate possible expression of the character of the organism, since they are directly related to the structure of the genetic material. In the case of the Arachnida, the only work concerned with the taxonomic aspects of silk composition is that of LUCAS et al. (1960). These workers determined the amino acid composition of 74 fibroins from a variety of silk-producing arthropods (including six spiders of three families) semi-quantitatively; 26 of these (including two spiders) were determined quantitatively. Some of the differences found (e.g. *Nephila madagascariensis* and *N. senegalensis*) appear to be based on testing silk from different glands rather than on genuine differences between species. Although the work of LUCAS and co-workers makes an interesting start, the problem of obtaining enough data in the case of a class that contains some 40,000 species is a formidable one. However, there is no doubt that a comparison of the evolution of the composition of silk would make an interesting comparison with the speculations of KASTON (1964) on the evolution of web types.

Amino acid composition of individual fibroins of Araneus diadematus. The amino acid composition of the various types of fibroin of *Araneus diadematus* is given in Table 3. The determinations were made by paper chromatography (PEAKALL, 1964b, and unpublished). Values for the total web, pulled thread, and cocoon have been determined by FISCHER and BRANDER (1960); and for the cocoon and dragline by LUCAS (1964). There are some discrepancies between a few of the values quoted, especially for alanine and proline, but these are not serious. The values given in Table 3 cover all the types of glands and are comparable to each other. It will be noted that the composition of the frame thread, pulled thread, and the material used in the web up to the completing of the scaffolding are essentially identical. The total web, which contains material from both the ampullate and aggregate glands, varies from the scaffolding threads in having a much lower value for glutamic acid and a much higher value for proline and isoleucine. The egg cocoon fibroin has a high serine and a low glycine content as compared to the web proteins. The swathing silk also has a higher serine and lower glycine content than the web proteins, but not to such a marked extent as the cocoon proteins. The attachment discs are rather similar to, but not identical with, the material from the ampullate glands.

Molecular weight. BRAUNITZER and WOLFF (1955) have found that the silk of the spider *Nephila madagascariensis* has a molecular weight of 30,000 in the liquid form and of 200—300,000 in the drawn fibroin. These measurements are in agreement with the finding of SMITH (1966) that the molecular weight of individual polypeptide chains do not normally exceed 20—30,000.

Table 3. *Amino acid composition (g/100 g) of the silk fibroins of Araneus diadematus*

	Total web Ampullate Aggregate	Frame thread Ampullate	Pulled thread Ampullate	Scaffolding of web Ampullate	Egg cocoon Cylindrical	Swathing silk Aciniform	Attachment discs Piriform
Alanine	27.3	32.8	32.7	33.4	25.4	25.3	29.3
Glycine	20.1	23.8	24.3	24.3	11.9	15.3	24.7
Serine	5.3	5.8	6.3	6.4	18.7	11.7	5.3
Total of Short Side-Chain.	52.7	62.4	63.3	64.1	56.0	52.3	59.3
Arginine	3.4	4.2	3.2	3.6	5.0	3.8	4.8
Aspartic	1.8	1.3	1.3	1.3	2.3	2.1	1.8
Glutamic	9.1	17.7	17.8	17.8	13.6	11.0	15.3
Isoleucine	5.7	2.1	1.7	2.4	5.0	4.8	2.1
Leucine	2.9	1.8	2.1	1.5	3.5	3.2	2.3
Lysine	3.2	2.1	1.8	1.9	1.3	2.2	2.1
Phenylalanine	1.1	n.d.	n.d.	n.d.	n.d.	n.d.	n.d.
Proline	12.8	3.1	3.1	2.3	3.8	10.2	4.7
Threonine	2.1	1.8	2.1	1.7	3.1	1.9	2.1
Tyrosine	1.7	2.4	1.8	2.1	3.7	3.7	2.6
Valine	3.5	1.1	1.9	1.3	2.7	4.8	2.9

Structure and conformation of fibroins. The crystalline portions of all of the filamentous silk proteins examined *(Bombycidae, Saturniidae,* and *Araneae)* show antiparallel pleated sheets of polypeptide chains packed into orthorhombic unit cells. All have repeated distances of 6.95 Å along the fiber axis and an equatorial backbone spacing of 9.44 Å. Thus, two dimensions of the unit cell are identical for all silks examined. The third dimension, which is the equatorial side-chain spacing, varies from 9.3 Å in the silk worm *(Bombyx mori)* to 15.7 Å in the spider *(Nephila senegalensis)* (WARWICKER, 1960). The values for *Araneus diadematus* are 15.0 Å for the drag-line and 15.6 Å for the cocoon filament (LUCAS, 1964). DOBB et al. (1967) have examined the structure of the silk of *Bombyx mori* by electron microscopy and X-ray diffraction. They find that the crystalline material is in the form of ribbon-like filaments of considerable length parallel to the fiber axis and of lateral dimensions approximately 20×60 Å.

Ordering of the thread on extrusion. The fundamental irreversible change from a water-soluble material to an insoluble fibrous thread is not completely understood. Basically it involves the reorientation of the molecules of the polypeptide from the soluble α-form with predominantly intramolecular hydrogen bonding to an insoluble, intermolecularly bonded β-pleated sheet structure of the fiber. RAMSDEN (1938) produced evidence that the transformation from soluble to insoluble forms could be brought about by shearing forces. He squashed the fibrinogen paste obtained from the larvae of *Bombyx* between flat, transparent plates and observed that an irregularly striated, semi-opaque, doubly refractive membrane was formed. RAMSDEN's experiments were mechanical in nature and gave no suggestion of the biochemical mechanism involved. Infra-red spectra studies have been made on the silk of *Bombyx mori* by AMBROSE et al. (1951) and LENORMANT (1956) who found that the material direct from the gland was largely amorphous, but extensively hydrogen bonded, largely intramolecularly. On pulling, the material is readily converted to the β-form. However, some α-fibroin is always present and complete polymorphic purity of the

Table 4. *Physical properties of some fibers* (LUCAS, 1964)

	Tenacity (g/denier)	Denier (weight in g of 9 km)	Extension at break
Araneus diadematus (drag-line)	7.8	0.07	31
(cocoon)	2.2	0.7	46
Bombyx mori (degummed)	3.7	1.0	16
Nylon (high tenacity)	8.7	15.0	16

β-form has not been found. There appears to be no comparable work on the silk of any spider. WARWICKER (1960, 1961) has published X-ray diffraction photographs on a wide variety of fibroins including those of *Araneus diadematus* and *Nephila senegalensis*. He concludes that, while no current theory is completely adequate to account for the fine structure of the fibroins, the data is best fitted by "random packing of anti-parallel pleated sheets of polypeptide chains".

Physical properties of fibroins. Some of the physical properties of the fibroins, taken from LUCAS (1964), are summarized in Table 4.

It can be seen from the table that the strength of the dragline is almost as good as that of high-tenacity nylon and its extensibility is considerably better. Since the drag-line is used to support the spider as it falls, both of these properties are of importance.

Detailed studies of the bearing of chemical composition on physical properties have not been made for the various types of fibroins of the spiders. It would be of considerable interest to study the relationship of the chemical composition to the physical properties for the material produced by each type of gland. LUCAS et al. (1955) have performed some studies along these lines for other silk fibroins. These workers investigated three groups of fibers typified by the Anaphe moth *(Anaphe sp.)*, the silk-worm *(Bombyx sp.)* and the Tussah moth *(Antheraea sp.)*. The fibroins were all cocoon fibroins and the Anaphe group contained the only spider material, the cocoon fibroins of *Nephila madagascariensis*. The value of the ratio of the long side-chain amino acids to short side-chain acids ($\times 100$) varied from 3.2 to 9.3 for the Anaphe group, was 14.5 for *Bombyx mori*, and varied from 22.1 to 30.0 for the Tussah group. Stress-strain diagrams showed that the presence of increasing numbers of amino acids with bulky side-chains led to increasing ease of folding or other distortion of the long molecular chains and thus to increasing extension of the fibers when they were stressed. The stress-strain diagrams of the Anaphe group were almost linear up to the point of rupture. The Tussah group of fibers showed relatively high extension after an initial resistance to extension of low load. This effect was considered to be due to weak cross-bonds that break at a certain stress in the amorphous sections of the molecule. The fibers of Bombyx were intermediate in character. The moisture regained by the fibroins did not vary much, even though the proportion of hydrophilic side-chains varies considerably. It is considered that the ability of the peptide linkage to bind water outweighs the effects due to the free hydrophilic groups. The apparent density of the fibers in water is greater than that measured in benzene. The increase is less for the Anaphe fibers and the fibers of Bombyx than for the Tussah fibers. This change is related to the ease with which water can penetrate and therefore, to the proportion of bulky side-chains.

Control of Drag-Line Spinning

WILSON (1962 a, b, 1967) has studied the regulatory valves of the ampullate glands of *Araneus diadematus* and their importance in the control of drag-line spinning. These valves are located at the base of the spinneret, about 700—1000 μ from the tip of the spigot. He has also studied the muscles associated with the valves in considerable detail. He considers that the valves can be regulated all the way from the closed to the fully open position. However, from our innumerable experiments on reeling silk from the spider, we have not observed any ability of the spider to interrupt the silk except by use of a leg. Whether the spider cannot or merely does not block this reeling of silk is a matter of conjecture.

WILSON considers that dragline spinning depends on three factors: — first, the body pressure of the spider forcing liquid silk up the duct from the gland; second, the control valve regulating the flow of liquid silk; and third, the tension in the silk thread aligning the molecules. It is considered that the intra-abdominal pressure moves the fluid silk up the duct and thus keeps the supply available for spinning. As there are no muscles around the silk gland itself, this pressure must be due to internal hydrostatic pressure. Although the musculature of the abdomen is limited, the frame-work arrangement (see Fig. 2) does appear to be capable of readily decreasing the abdominal volume and therefore, increasing the pressure. WILSON has measured the difference in pressure between active and resting spiders and finds it to be 3—4 cm of mercury.

The control valve is responsible for determining the thickness of the thread. The diameter of the main duct of the ampullate gland is some ten times larger than the thickness of the thread. The silk is in the form of a thread of final diameter between the control valve and the flexible cuticle tip of the spigot (see diagram 8 b, WILSON, 1962 b). The importance of being able to alter this parameter of the web has been demonstrated in starvation experiments (see page 53), in the administration of physostigmine (see page 65), and in the addition of weights to the backs of the spider (see page 49). Under all these conditions the diameter of the thread used in web-building is altered.

We have found that the time spent in the center of the web between the laying down of consecutive radii is long, compared with the time taken to lay down the radii (see page 79). It is possible that the pauses between each radius-placement may be necessitated by the need to pump silk into the duct. Although such a conclusion is difficult to prove, it can be calculated from the data given in Table 2 that the amount of material in the duct is of the same order of magnitude as that used in the construction of a single radius. Such a theory would account not only for the long time spent at the center, but also for the fact that the duct is several times longer than is necessary, if it has the sole function of connecting the main part of the gland with the spigot. Another possible function for a long duct may lie in the ordering process of the fibroin and/or water removal (see pages 13, 9). These possibilities are not mutually exclusive.

B. Regulation of Protein Synthesis in Silk Glands

In the previous chapter it was shown that each set of silk glands has the sole function of producing a single structural protein. In this chapter the regulation of production of this protein will be considered. Two modes of stimulation have been demonstrated, one cholinergic and one related to the emptying of the gland. Both of these mechanisms are capable of stimulating the gland to a similar extent and the effects are not additive. Only the cholinergic stimulation can be blocked by atropine.

The process of stimulation can be divided into two main sections and these apply to both the cholinergic and emptying modes of stimulation. The first process is the release of pre-formed protein from the epithelial cells into the lumen of the gland. This is followed by an increase in the rate of synthesis of new protein. It is possible to block the synthesis of new protein without affecting the initial secretion of pre-formed protein.

The over-all pattern is similar for the two modes of stimulation but the cholinergic stimulation produces increased levels of protein synthesis over a shorter time period. We consider that it is possible that the cholinergic mechanism acts as a fine control superimposed on the other mechanism which is dependent on the level of protein stored in the gland.

The question of comparison of silk production in spiders and in the silk worm *(Bombyx mori)* is often raised. Although there is considerable similarity in the chemical composition of the fibroin, the similarity between the silk producing glands of *Bombyx* and *Araneus* is rather superficial. *Bombyx* produces its silken cocoon only once in a lifetime and the process takes several weeks, the spider produces a web a day in less than an hour and can produce even more frequently than this. Thus the regulating mechanism is completely different. For example, the maximum production of messenger RNA in *Bombyx*, as measured by the incorporation of C^{14}-orotate, is on the 3rd day of the 5th instar when protein production is low. When protein synthesis reaches its peak 3 days later, the rate of m-RNA production cannot be detected (MIURA et al., 1965). In *Araneus* the entire cycle is completed within 8 hours.

Methods

The following methods were used in the investigation of silk synthesis:

1. Reeling or pulling thread from the living spider. Since the spider lays down a silk thread as it moves, it is possible to reel silk from the animal by attaching the end of the thread to a glass rod which is turned by a small motor. The spider is held by hand until all the thread has been reeled onto the glass rod, a process that takes five to ten minutes. Histological studies have shown that the ampullate glands are emptied by this process. The procedure, which must be the simplest known extraction of a pure protein from a biological system, can be done without damage to the spider. Thus the method can be used to study the production of silk in an individual spider over a period of time. This reeling procedure was formerly used commercially in Madagascar using the spider *Nephila madagascariensis*.

2. **Determination of nitrogen.** For the determination of nitrogen the silk is reeled onto a small piece of low nitrogen filter paper and this is digested in 0.2 ml of selenium sulfuric acid until colorless. Then 0.4 ml of 15 N NaOH, 3 drops of 2% gum ghatti, 3 ml of NESSLER's reagent are added and the volume made up to 15 ml with distilled water. The light absorption is measured on a spectrophotometer. Determinations can be made down to 5 µg of nitrogen.

3. **Autoradiographic studies.** In this technique the position of radioactively labeled material in fixed sections of tissue is revealed by the development of photographic emulsion layered above the section by the emitted radiation. The spiders are fixed for 5—6 days in buffered formaldehyde, double embedded in collodion and paraffin and cut on a microtome at 6 µ. After the removal of the paraffin the slides are dipped in Ilford L 4 photographic emulsion, dried and stored in the dark for 1—4 weeks. The emulsion is then developed and the position of the radioactive isotope is shown by black dots where the β-rays have affected the emulsion. This method gives information as to the localization of the isotope within the cell and is valuable in the tracing of the time sequence of events. It does not give information on the chemical composition of the material containing the label. Quantitation is possible by counting the number of dots per unit area but is generally less satisfactory than counting radioactivity in a scintillation counter.

4. **Scintillation counter.** The total incorporation of radioactively labeled material is measured by dissolving the silk or gland in N NaOH and adding to BRAY's solution. The activity is measured in a liquid scintillation counter and the degree of quenching calculated against an internal standard. These measurements give the total amount of labeled material taken up in a given experiment, but give little information on cellular localization.

Cycle of Events Following Cholinergic Stimulation

The sequence of events in the epithelial cells following stimulation will now be considered in some detail. Histological studies have been made with light and electron microscope, including autoradiographs. Chemical and spectrophotometric determinations have been performed on nucleic acids and protein concentrations.

If a single dose of acetylcholine (1 mg/kg) is injected into the abdomen of the spider, stimulation of protein synthesis in the silk gland is observed. Autoradiographs following the injection of C^{14}-acetylcholine (both acetyl-C^{14} & choline-C^{14}) show that the acetylcholine is localized on the outer membrane and does not enter the cell (PEAKALL, 1968b). An autoradiograph of C^{14}-acetylcholine made from a section of a spider killed 5 minutes after injection is shown in Fig. 4.

It was found that 1 mg/kg was a maximal dose as far as surface labeling was concerned. Experiments on isolated glands with atropine over the concentration range 10^{-8} to 10^{-4} M showed a reduction in the number of counts per unit length of membrane. Diisopropylfluorophosphate(DFP) had no effect on the extent of labeling on the membrane at low concentrations ($10^{-8}-10^{-10}$ M), but significant reductions occurred at 10^{-6} M. The effects of atropine and DFP were not found to be additive. The amount of labeled material localized along the membrane was greater on the tail portion of the gland than on the sac portion. The actual counts, based on 50 counts from each of three spiders, were 12.4 ± 3.2 per μ for the tail compared to 7.4 ± 3.2

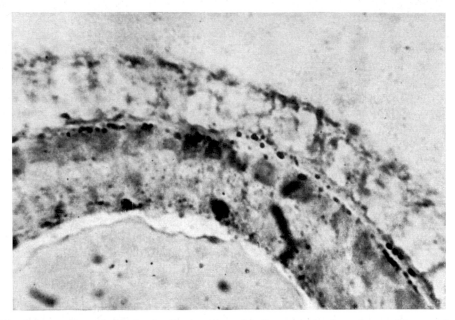

Fig. 4. Autoradiograph showing localization of acetylcholine on the basal membrane of the ampullate gland. Acetylcholine (methyl C^{14}) chloride was injected into the spider, the gland fixed in buffered formaldehyde 5 minutes later, double embedded in collodion and wax, sectioned at 6 μ. The slide was exposed for 14 days using Ilford L4 emulsion before development, magnification $\times 970$, stained with azure B

for the sac. This difference is in agreement with the reduced rate of synthesis of protein in the sac portion of the gland.

The sequence of events following stimulation can be divided into two main divisions: secretion into the lumen of preformed protein and the synthesis of new protein. This division is clearly shown by pretreatment with an inhibitor of protein synthesis followed by stimulation. The inhibitors used were actinomycin D and puromycin; control experiments showed that the dose of inhibitors used was not lethal within a few days. In both cases the secretion of the preformed protein occurs normally but the synthesis of new protein is blocked. The large droplets of protein formed following stimulation are clearly seen in the electron microscope and are shown in Fig. 5. Autoradiographs of spiders which were injected with labeled alanine along with the acetylcholine showed that this protein was unlabeled, which confirms that these droplets consist of protein synthesized before stimulation.

The appearance of the "resting gland" in the electron microscope is shown in Fig. 5 and diagramatically in Fig. 6a. The term "resting" is used in contrast to

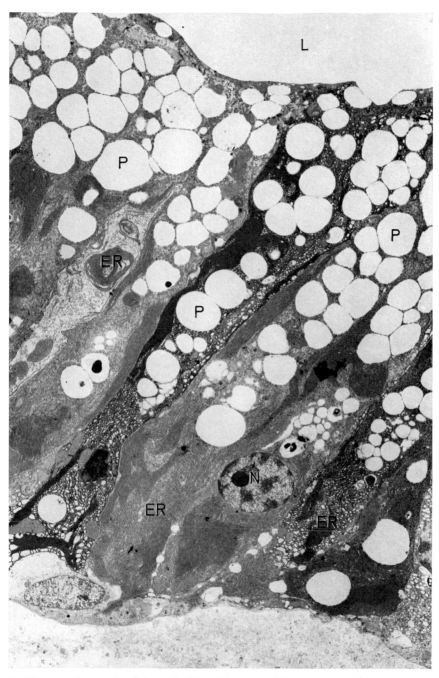

Fig. 5. Electron micrograph of the wall of the tail region of the ampullate gland showing protein droplets (*P*), nucleus (*N*), rough endoplasmic reticulum (*ER*). The lumen (*L*) of the gland contains a substance of the same electron density as the droplets. Fixed in 3% glutaraldehyde in Sorensen's phosphate buffer at pH 7.4; post fixed in 1% OS O$_4$ in same buffer. Embedded in Epon and sectioned c̄ diamond knife, stained with uranyl acetate and lead citrate. Micrograph taken on Philips 300 electron microscope at 1300 × and enlarged photographically to 4,100 ×, by Allen Bell

2*

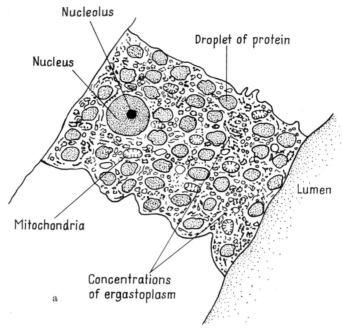

Fig. 6. Diagrammatic representation of the structural changes occurring in the epithelial cells of the ampullate gland with stimulation. a) Appearance of cell before stimulation. b) 5—10 minutes after stimulation. c) 20 minutes after stimulation. d) An hour after stimulation

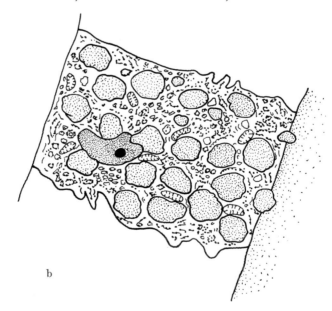

"stimulated" and should not be taken to mean complete inactivity. Studies of the incorporation of labeled amino acids showed that synthesis of protein occurred at all times although often at a reduced rate. In the resting state droplets of preformed protein are seen throughout the cytoplasm, the mitochondria are distinct and are

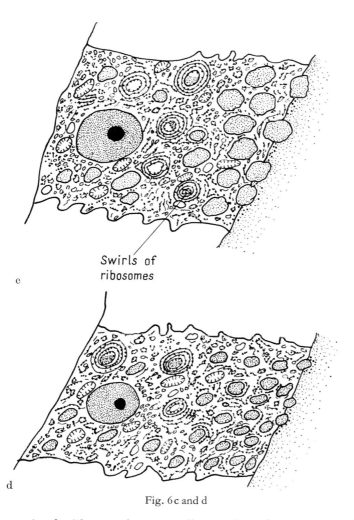

Swirls of
ribosomes

c

d

Fig. 6c and d

not found associated with any other organelles; nuclei and nucleoli are fairly small.
No Golgi apparatus has been found at this, or any other, stage (Bell and Peakall,
in press). Some discharge of protein into the lumen has been seen at this stage
although to a much lesser extent than occurs after stimulation. This is in keeping
with the known relative activities of the cells at these times.

The appearance of the cell 5—10 minutes after stimulation is shown diagram-
matically in Fig. 6b. Large droplets of preformed protein, visible in the light microscope, are seen throughout the cytoplasm. The nucleus is often deformed in shape
by the droplets.

20 minutes after stimulation (Fig. 6c) the discharge of these droplets into the
lumen is well underway. Swirls of ribosomes are seen in the cytoplasm; the nucleus
and nucleolus are increased in size. The swirls of ribosomes fill with protein and the
membranes surrounding them are partly rough and partly smooth. There is considerable ribosomal debris at the edge of the droplets. The secretion of the droplets
into the lumen occurs without going through a Golgi apparatus. If the Golgi

apparatus exists at all in these glands, it cannot be common enough to be involved in the production of silk fibroin. The complex membrane structures seen at this stage may well be involved in the 'reverse phagocytosis' process. The general appearance of the transport of droplets in these cells supports the view of KAR- NOVSKY (1962) that the formation and breaking of membrane links underlies the active passage of substances during secretion.

The discharge of protein into the lumen causes considerable damage to the cell structure. There is a concentration of protein droplets discharging into the lumen and also of ribosomal material along the epithelial/lumen boundary. This can be seen both in the electron microscope and in the increased concentration of RNA in the luminal end of the epithelial cell as measured with a microspectrophotometer. Electron microscope photographs show the cytoplasm to be rather empty in the region around the nucleus. These photographs also suggest that some ribosomal material is lost into the lumen at the time of discharge. Such a loss of ribosomal material has been confirmed in two ways. First, spiders fed with orotic-6-T acid have some radioactivity in their silk. It is possible that this activity arises from the incorporation of tritium into some other compound than RNA. Second, autoradio- graphs of stimulated glands pretreated with orotic-6-T acid shows light, but above background, labeling in the lumen of the gland. Further, this labeling is not found if the section is pretreated with RNase, thus demonstrating the specific nature of the labeling. The importance, or otherwise, of traces of ribosomal material in the silk is unknown.

After an hour (Fig. 6d) the appearance of the cell is similar to the resting state, but light microscope autoradiographic studies show that synthesis of new protein and its discharge into the lumen is still continuing at a rapid rate.

The profound movement of ribosomes around the cell have also been demon- strated by means of light microscope autoradiographs. In one experiment a half hour pulse of orotic-6-T acid was given at the time of stimulation. The glands were re- stimulated by injecting acetylcholine 24 hours later. The spiders were killed at given time intervals (0, 15, 30, 60 and 120 minutes) after stimulation. At zero time there was noticeable localization in the nucleus, although the cytoplasmic activity was considerable. 15 minutes after stimulation the activity in the nucleus was almost zero but there was considerable activity in the cytoplasm near the nucleus. This suggests movement of mRNA and/or newly synthesized rRNA from the nucleus into the cytoplasm. 30 minutes after stimulation the localization is largely at the luminal edge of the epithelium. This is in agreement with electron microscope and microspectrophotometric studies mentioned above. After an hour the localization has largely disappeared and no localization is seen after this time. If the cold orotic acid was not given until the second stimulation (i.e. a pulse of 24 hours) then vir- tually no localization was observed at zero time. However, the localization at the luminal edge of the epithelium 30 minutes after stimulation was seen.

Experiments using actinomycin D show that the onset of new synthesis occurs within a few minutes of stimulation. Actinomycin acts by blocking the DNA- dependent synthesis of RNA (HURWITZ et al., 1962), and thus stops protein synthesis at one of its earliest stages, the production of messenger RNA. Labeled amino acids were given with the acetylcholine, followed at various time intervals by actinomycin. The spiders were killed an hour after stimulation and the glands examined by autoradiography and by determining the amount of incorporation into protein using a scintillation counter. If actinomycin was given with the acetylcholine or within 3 minutes of the acetylcholine, then protein synthesis was found essentially completely blocked. If actinomycin was given 4 minutes or later, it had little effect on the amount of labeled amino acids incorporated in the one hour period. These results show that the DNA-dependent synthesis of RNA occurs within the first 3 minutes following stimulation (PEAKALL, 1968, in press).

Experiments have been carried out to calculate the half-life of the mRNA. In these experiments actinomycin was given 15 minutes after stimulation — i.e. after the first burst of mRNA synthesis — and the amount of protein synthesized over various time periods was measured by the incorporation of labeled amino acids. It is found that messenger RNA is relatively stable: it has a half-life of a few hours. In this respect it resembles mammalian systems rather than the short-lived bacterial mRNA (COHEN, 1966) or the silk-worm (where the synthesis of messenger RNA occurs several days before it is used, MIURA et al., 1965).

Autoradiographs using labeled alanine given at the time of stimulation show significant increases over control in the time period 30—60 minutes after injection of acetylcholine. Electron microscope photographs suggest that synthesis of new protein may occur earlier than this, viz. 15—20 minutes post-stimulation. This difference has been resolved by pulsing-in the labeled alanine 15 minutes before stimulation — the cold alanine is given with the acetylcholine. In this way allowance is made for the movement of amino acids from the body fluids into the amino acid pool of the epithelial cell. Little activity is seen in the lumen until an hour after stimulation. Thus the discharge of protein seen in the electron microscope 20—30 minutes after stimulation must consist of preformed protein. Heavy discharge of labeled material into the lumen is seen in the 1—3 hour period after stimulation. By four hours post-stimulation the discharge of labeled material has decreased and the activity in the epithelial cells instead of being general is now concentrated in small droplets. The appearance of the cell then remains essentially unaltered until the gland is re-stimulated. Then these labeled droplets are rapidly (in 20—30 minutes) discharged into the lumen. It appears that after the formation of these droplets the rate of protein synthesis falls to a low level.

The mechanism by which cholinergic stimulation causes increased RNA and protein synthesis is not known. Acetylcholine could conceivably function as a component of a primary signal for new RNA and protein bio-synthesis. More likely, the primary effect of the cholinergic stimulus is to increase the secretory rate of the gland. RNA and protein synthesis may then be stimulated by a signal that is generated by the release of stored protein. The secretion of enzymes in the pancreas is increased by cholinergic agents (HOKIN and HOKIN, 1961) and in the pancreas of the pigeon a link between secretion and increased rate of synthesis of protein has been demonstrated (PEAKALL, 1967b). It is possible that this cholinergic mechanism has wide application.

Relationship to the "Natural" Cycle

Most of the experiments on the 'natural' cycle have been carried out on spiders whose silk glands have been emptied by reeling out the silk. Histological evidence shows that both web-building and reeling of the silk effectively empty the ampullate gland. However, the possibility of differences between the two methods cannot be ruled out. The reeling procedure has the advantage that it can be done at any time of the day and most seasons of the year[1]. The spiders used in these experiments are housed in small (250 ml) jars and do not build webs under these conditions, although some limited amount of silk is used by the spider as it moves about in the jar.

Basically there is considerable similarity between the events following cholinergic stimulation and those following the emptying of the gland. The same movement of the preformed protein into the lumen is seen followed by the rapid synthesis of new protein. The rate of synthesis of new protein, as measured by the incorporation of labeled amino acids, over the period of 1 to 2 hours after stimulation, is the same. The time of onset of new synthesis following stimulation was measured by reeling thread out of the spider as rapidly as possible for 2 minutes and then injecting actinomycin at various time intervals afterwards. It was found that synthesis of new protein could be blocked if given within 3 minutes of the end of the reeling process. Thus the time of onset of new synthesis is similar to that found for the cholinergic stimulation, although detailed comparison is difficult since it depends upon when in the reeling process the stimulation is assumed to start. If the start of emptying is taken as time zero, then the onset of new synthesis is slower (by approximately 2 minutes) in the case of the 'natural' cycle (summary see Table 5).

Table 5. *Two ways of stimulating silk synthesis in ampullate glands*

(1a)	Acetylcholine binds with receptor on outer (basal) membrane of gland	Signal across outer epithelial membrane causes release of presynthesized fibroin	Atropine sensitive Puromycin insensitive
(1b)	Emptying of the gland	Signal across inner (luminal) epithelial membrane causes release of presynthesized fibroin	Atropine insensitive Puromycin insensitive
(2)	Signal after release of presynthesized fibroin from epithelium	Increase in nuclei size; change in RNA concentration; increased incorporation of amino acid	Puromycin sensitive

There is, however, a distinct difference in the duration of accelerated synthesis of new protein. A 15 minute pulse of labeled alanine was given immediately after the reeling process was completed. In these experiments the reeling was carried on until no more thread could be obtained. After 1 hour it was seen that a moderate amount of new labeled protein had been formed, but little had reached the lumen. Moderate activity of the silk in the lumen was noted by 2 hours and the discharge of labeled material into the lumen was still heavy at 4 hours. The discharge activity had almost ceased after 8 hours and now the main characteristic was the appearance of heavy labeled storage droplets in the epithelium. This was similar to the appearance seen in the cholinergically stimulated gland 4 hours after stimulation. The appearance of the gland at 16 hours was essentially similar to that at 8 hours. The labeled storage droplets could be readily discharged by reemptying the gland.

The turn-over of labeled nuclear proteins in relation to the secretion-synthesis cycle that occurs in the epithelial cells following web-building has been examined

---

[1] In late winter the spiders are sluggish and it is not possible to pull out thread. At this time of year the rate of incorporation of alanine into silk protein is reduced to a quarter of its normal value

(PEAKALL and CAMERON, 1968). Previous experiments to study the conservation of nuclear proteins (i.e. PRESCOTT and BENDER, 1963; GOLDSTEIN and PRESCOTT, 1967) have been carried out either by removal and transplantation of the nucleus from its original cytoplasm or by successively cutting-off cytoplasmic material. Using the silk gland it is possible to study the conservation of nuclear proteins in a differentiated metazoan cell system where most of the cytoplasmic protein is removed during the normal cell cycle. Labeled lysine was given to the spider both as a short pulse and by feeding labeled lysine to the spider daily for 5 days. Experiments were also carried out with labeled thymidine to discover what percentage of cells were synthesizing DNA. It was found that there was a virtually complete loss of lysine labeled nuclear material in all of the spinning gland cells after the spider has spun three to five webs although those cells with labeled DNA (2—5%) appear to have complete metabolic integrity of their DNA during the same web-building activity. It is clear that the lysine-labeled nuclear material does not have the same order of stability as DNA, even in those cells which were in nuclear DNA synthesis at the time of isotope administration. The suggestion that the lysine-labeled material plays an active role in the regulation of protein synthesis is supported by preliminary experiments which show that the labeled material leaves the nucleus within half-an-hour after web-building.

Regulation of Silk Glands Other than Ampullate Glands

The regulation of the aggregate glands has not been studied in any detail. However, examination of autoradiographs shows that the aggregate gland is stimulated by cholinergic agents and labeled acetylcholine is found localized on the surface of these glands. It appears that the regulating mechanism parallels that of the ampullate gland, as would be reasonable in view of the function of these glands.

No studies of the regulation of the aciniform and piriform glands have been made but some localization of labeled acetylcholine has been observed on the outer membranes.

The regulation of the cylindrical glands, those used to produce the silk for the cocoon, is completely different from the other glands. No uptake of labeled amino acids is observed within 8 hours, a period of time in which both aggregate and ampullate glands show heavy incorporation. Injection of acetylcholine does not cause the take-up of amino acids nor does labeled acetylcholine show any localization on the surface of the gland. After the silk has been used in the manufacture of the cocoon the gland does not refill, at least within 30 days (PEAKALL, 1965b). Thus it seems clear that this gland has neither the cholinergic nor the local feedback mechanism which have been demonstrated in the ampullate gland. In view of the fact that the sole function of these glands is to produce silk for the cocoon it is likely that silk production in them is regulated by hormones, but this possibility has not been examined.

Amounts of Protein in Web-Building

WITT has found that the nitrogen content of webs is increased by physostigmine and decreased by atropine, psilocybin and diazepam. The changes obtained are summarized in Table 6 and are considered in more detail under the individual drugs in

Chapter 4. An increase in size of webs has been found following the administration of amphetamine, but nitrogen values have not been obtained. In addition there is an increase in the total amount of protein used in web construction by spiders with a lead weight (amounting to 30% of their body weight) attached to the back of the spider (Table 6).

Table 6. *Nitrogen content of webs following various treatments*

Pretreatment	Dose and time after per os administration		Nitrogen content, μg N		Reference
			Control	Experimental	
Physostigmine	1 mg/kg	36 hours	36	49	WITT (1963)
Atropine	4 mg/kg	16 hours	110	69	WITT (1962)
Diazepam	100 mg/kg	36 hours	35	20	REED and WITT (1968)
Psilocybin	150 mg/kg	24 hours	112	65	CHRISTIANSEN et al. (1962)
Weight (equivalent to 30% body weight glued on back 24 hours before)			29	39	CHRISTIANSEN et al. (1962)

The data obtained from experiments in which silk is drawn from the gland differ from data obtained from web nitrogen in two respects. First, the only gland affected by the reeling process is the ampullate gland whereas in web experiments materials from ampullate and aggregate glands are measured. Second, once the reeling operation is started, the spider has no means of stopping the thread from being reeled out. Thus it is possible that a drug will decrease the amount of nitrogen in the web, i.e. reduce web size, by factors that do not affect silk production and therefore do not affect the amount of silk that can be reeled out. This has been demonstrated for the tranquilizer diazepam and these results are discussed in Chapter 4. Thus comparison between effects on amounts of nitrogen in the web and that reeled out can give valuable information on the mode of action of the drug in question.

There are four experiments which independently suggest that the spider has information on the amount of silk available for web building. Since function of the web as a food snare is reduced to the extent that it is not completed, the value of this information to the spider is obvious. The experiments, all considered in more detail elsewhere in this book, are (1) starvation experiments in which the spider makes a smaller, wider meshed web (see page 53); (2) addition of weights to the back of the spider when a thicker, but shorter thread is used in a full sized web (see page 49); (3) oral administration of physostigmine when more protein is used in a larger web, especially in the sticky spiral (see page 65); and (4) the burning out of the last radii (see page 77): in this case the spider replaces the radii for a while, but finally ignores the oversized angle and goes on to building the spiral. This experiment is particularly interesting since it indicates that the spider may have a continuous feed-back of information on the silk supply: adjustment is made to a situation that arises during the actual building of the web.

The question is, of course, how this feed-back of information is mediated. Although the emptying of the gland causes rapid synthesis of new protein in the epithelial cells, it does not appear to be possible that any of the changes in these cells can be correlated to the amount of silk remaining in the lumen. Serial sections of the abdomen of the spider were examined for a mechanoreceptor which could act as a transducer.

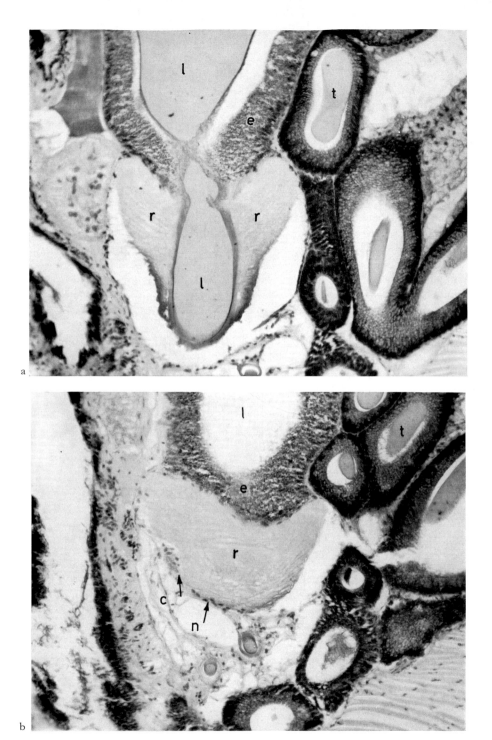

Fig. 7. Cross-sectional views of ampullate gland showing a structure considered to be a receptor sensitive to the amount of silk stored in the gland. Lumen (*l*); epithelium (*e*); receptor (*r*); tail of ampullate gland (*t*); nuclei at edge of receptor (*n*); connective tissue (*c*). Magnification ×140; stained with hematoxylin and eosin

The appearance of a structure on the ampullate gland that may be the receptor is shown in Fig. 7. In Fig. 7a the 'receptor' is in contact with the silk in the lumen of the gland; in this section, the segment of the lumen in which the receptor occurs is closed off from the main part of the lumen, while in the next section (not shown) it is clearly open (PEAKALL, 1968a). The appearance of the receptor is very different from the normal epithelium. The epithelium takes up a good deal of the dye hematoxylin, whereas the receptor is largely stained with eosin. The receptor has a somewhat striated appearance. Cross-sections of the coiled tail of the ampullate gland are also in this photograph. Fig. 7b is a photograph of another section of the same spider, 100 μ away from the section in Fig. 7a. At this stage the structure is not in contact with the lumen of the gland, but the peripheral nuclei of the receptor cells and the associated fibers can be seen. Staining methods for the nerve fibers are less than adequate in the abdomen of the spider. However, it has been possible to show, albeit poorly, using silver staining, that fine nerve fibers are present in the connective tissue. Rigorous proof of the function of the structure has not been obtained. It is a reasonable conjecture that the structure is a transducer which converts pressure or tension into nerve impulses. The structure is found in the storage portion of the gland where the silk is known to be in liquid form. The striated portion could signal a change in the pressure or tension, information which could be transmitted via nerve fibers to the central nervous system. In general appearance it is not too dissimilar from the Pacinian corpuscles, a structure in mammalian skin and muscles which converts pressure into nerve impulses (LOEWENSTEIN, 1960).

III. Specificity of the Web

There exist a number of observations which indicate that the pattern which an adult female *Araneus diadematus* Cl. spins is characteristic for the species. Comparison of webs of the closely related *Araneus diadematus* and *Araneus sericatus* (Figs. 1 and 8) shows a striking similarity in the patterns, while geometric webs of builders only remotely related construct patterns which look quite different. Efforts have been made to trace the history and relationship of species of spiders by means of web

Fig. 8. This web of an adult female *Araneus sericatus* Cl. was built and photographed under the same circumstances as the web in Fig. 1. Note similarities as well as differences in the two webs

resemblances under the assumption that a more "primitive", simple, irregular web was the forerunner of the more elaborate patterns. For a number of reasons such deductions seem at present precipitous, and other approaches such as determining the chemical similarity of silk proteins (see previous chapter) appear to provide a more secure basis for construction of a family tree of web-builders.

In comparing webs of different species, only those built by adult spiders or animals of comparable age should be used. For instance the number of web components changes with the age of an individual *Araneus diadematus*. Frequently such changes seem to depend on the growth of the animal which is measured as body weight and leg length. Such observations will be discussed later under "Changes

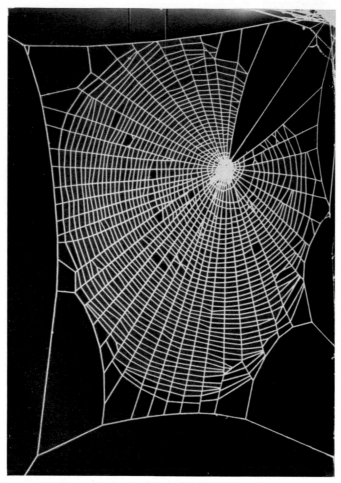

Fig. 9. This photograph of a sprayed web of an adult female *Zygiella-x-notata* Cl. shows the characteristic free sector and signal thread. The spider rests in the retreat in the right upper corner, front legs on signal thread

of Webs with Age". The pattern of the web made by a juvenile member of one species may resemble the web pattern produced by adults of another species. Moreover the technique of using the web for catching prey may be that of the other species. This has given rise to speculation that the phylogenetic history of the web is repeated in an individual's web-building sequence. However, there appears to be at present only one good example in favor of such a recapitulation.

An adult *Zygiella* builds webs with a free sector (Fig. 9). A signal thread runs from the hub through the free sector to the spider's hiding place; the "retreat" is always outside the web, under a leaf, for example. The assumption can be made that greater "skill" is necessary to construct such a web rather than one with an all-around spiral. Furthermore, it can be assumed that the *Zygiella* web is particularly advantageous for the spider's survival: the trapper no longer sits conspicuously in the middle of its trap, but waits well hidden in its remote retreat. In spite of the distance, the signal thread brings it into close tactile contact with the web center and permits it to rush quickly from the hiding place through the hub to the struggling prey.

In the laboratory as well as outdoors it can be observed that young *Zygiella* build *Araneus*-like all-around spirals and wait for the prey in the center, constructing the characteristic *Zygiella* web only later in life. PETRUSEWISZOWA (1938) and later MAYER (1953) have reported that the change from the round to the free-sector web appears at a certain time of life independent of building experience. *Zygiella* which were not permitted to build any web until they were adult, executed the first pattern appropriate to their age, with a free sector, independent of preceding experience. As far as I know, this example provides the only support existing for an ontogenetic web change which may point to a phylogenetic history of web-patterns. Even that is now in doubt following a report by PETERS (1967) on juvenile *Zygiella*, where the free sector seemed to be dependent on the nature of the frame rather than on the age. Other web changes, like those described by SZLEP (1961) in *Uloboridae* and in this book later, in *Araneus*, appear to be the result of body growth rather than ontogeny.

Examination of the geometric webs of different species shows that they all have some common feature (Figs. 1, 8, 9, 10a, 11, 12, 13). There is always a hub, densely meshed, with strong threads, radiating from there to the frame. Across the radii a more or less circular spiral is stretched; it can be sticky or not, but seems always to have a catching function. Even the web of *Hyptiotes* (Fig. 13), an *Uloborid* only very remotely related to *Araneus* (KASTON, 1966), shows this basic pattern. Such similarity in trap construction has at least two different explanations: it either indicates that the webs have developed from a common origin, or that this design is the simplest solution to the task of catching the greatest number of flying prey with as little effort and material as possible (WITT, 1965): adaptive convergence. The latter explanation would become even more plausible if it could be shown that very few, simple movement patterns and orientations are necessary for the construction of the geometric orb. One kind of demonstration which would lend support to such a possibility would be a successful computer simulation of web patterns which employed a minimum of assumptions about the spider's capacities for orienting and moving. Some first effort has been made by EBERHARD (1967) who has sought to show that an acceptable reproduction of the spiral of *Uloborus* can be made if certain aspects of web geometry are assumed somehow to guide the spider.

But can we see any evolutionary advantage in the development of so many different patterns when we keep the trend toward economy in nature in mind? LE GUELTE exchanged spiders so that they occupied webs of animals of similar size, but different species. Results of these experiments were reported by LE GUELTE (1967). Suffice it to say at present that spiders seemed able to catch and process house flies on strange webs with the speed and efficiency they showed in their own web. There may be subtle differences in speed of movement or amount of web

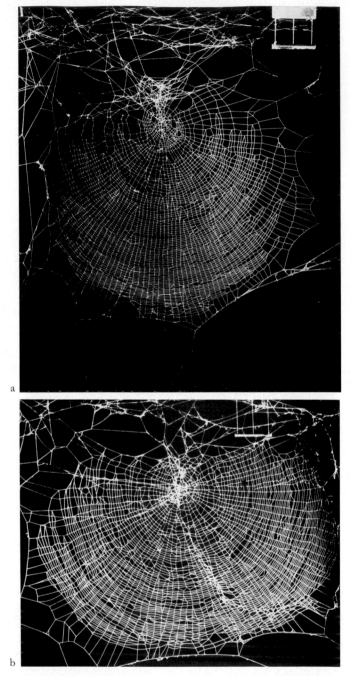

Fig. 10. a The web of an adult *Nephila clavipes* shows also radii, hub, and spiral, but characteristically is considerably more asymmetric than webs of *Araneus*. b The same spider built this web after 600 mg/kg d-amphetamine. Compare the irregular spiral spacings with those in *Zygiella* webs after methamphetamine (Fig. 20): the drug producesa similar variation on a different theme

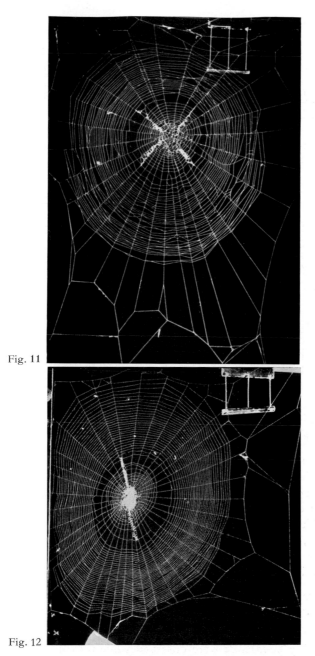

Fig. 11

Fig. 12

Fig. 11. A web of an adult female *Argiope argentata*. Note the hatched band or stabilimentum which
was added at the end of web-building; the function of this structure is unknown

Fig. 12. An adult female *Argiope aurantia* built this web in the laboratory; compare with web of
Argiope argentata (Fig. 11)

destruction in the catching process, but it is apparent that the animal is successful. This result led us to the hypothesis that different patterns of geometric webs are important for proper mating by providing an unambiguous signal for the approaching male about the nature of the female. Observations of the frequently hour-long process of approach of the sexes can make such an early warning system seem most economical. Clearly the subtleties of the form of the web do not limit its useful-ness for prey-catching by alien tenants; different patterns, as far as LE GUELTE's samples show, are equally workable.

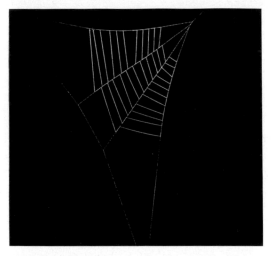

Fig. 13. Web of an adult *Hyptiotes cavatus* built in the laboratory frame. The animal holds the thread at the apex, releasing it only when prey has been caught. This web has to be rebuilt after each prey capture

Our present method of measuring and evaluating webs has intentionally neglected the individual web. Because web measurements were used to identify drug effects, and the evaluation of changes was statistical, measures of web geometry were chosen which were most common to all webs of one species. The common effect of the drug on all individuals' behavior was of interest rather than individual features. There is some indication, however, that an individual builds features into its daily pattern which distinguish it from its peers. Investigation of such features should be of interest in the elucidation of the genetic basis of behavior.

Three preliminary observations may illustrate the possibilities:

The shape of the frame which the spiders build around their webs seems to contain individual features which become particularly apparent under constant laboratory conditions. Two *Zygiella-x-notata* were observed while they lived together in one wooden frame with their retreats in the upper right and left corners respectively. The frame was partitioned in the middle by means of a wooden divider so that the webs did not touch (Fig. 14). Spider A in the right half of the frame built a new web every day with a point at the bottom, while spider B in the left half of the frame always built a nearly horizontal frame thread at the lower end. When one day spider A accidentally got into B's web, B was placed in A's web. Both accepted the

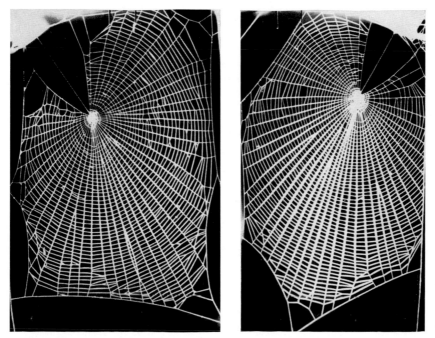

Fig. 14. These two webs of adult females *Zygiella-x-notata* were built in a subdivided frame with retreats in opposite corners. Note the horizontal frame thread at the bottom of the left web and the pointed shape at the bottom of the right web

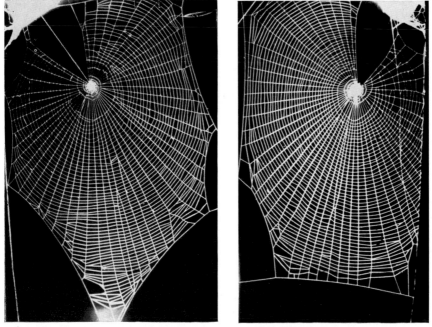

Fig. 15. The two spiders of Fig. 14 had been accidentally exchanged and each built another web in their "personal" pattern in mirror image

3*

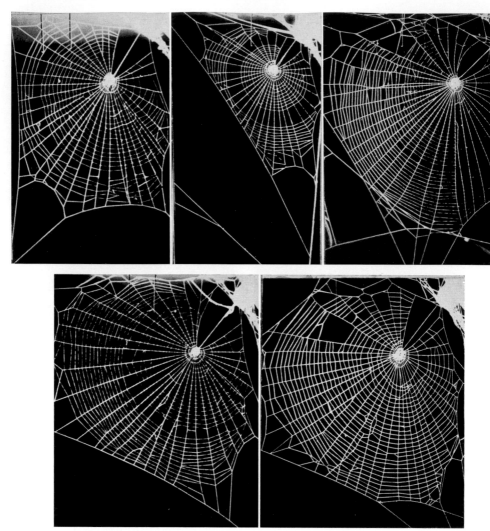

Fig. 16. One female *Zygiella-x-notata* built these webs on February 16 and 20, March 10, 12 and 14. Notice the peculiar free ring or parts of it in all webs; this appeared in the more than 100 webs photographed during this animal's life, not shown here

new webs and proceeded along the signal threads into the retreats. The webs were now eliminated, and Fig. 15 shows the new webs built on the following morning: the spider with the pointed web again built its characteristic pattern, starting now from the opposite corner, in mirror image. Spider B also preserved its characteristic web shape in its new surroundings, building its horizontal base thread in mirror image.

Another observation which sheds some light on individual web patterns concerns a *Zygiella-x-notata* which showed a remarkable abnormality in its webs: the animal always left one or more turns out in the middle of the spiral, but spaced the rest of the spiral normally (Fig. 16). This strange pattern was observed with variations for over a month in each daily web, and the figure shows only a small selection. In later life the open space slowly disappeared.

The third observation is left to the reader to make by examining the drawings of the outlines of webs made by one spider throughout more than three months of its life (Fig. 17). Again only a selection can be presented. No measure has as yet been developed to characterize the shape, and the reader is asked to simply compare their similarity to the different shapes of the preceding figures.

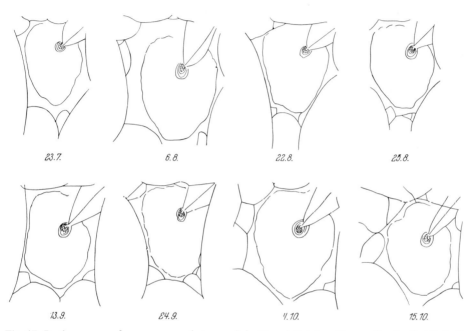

Fig. 17. In the course of one summer, between July 23 and October 15, one female *Zygiella* built these webs which appear rather similar in outline

There are many spiders which build no webs at all and others build webs with other than two dimensional geometric patterns. The large first group is of no interest for this discussion, but the others can be examined with respect to possible similarities to geometric orbs. A good example is provided by *Latrodectus*, recently thoroughly analyzed by Szlep (1965): The web of *Latrodectus pallidus*, for example, consists of three structurally, functionally and geographically separate parts. *1.* the *retreat* is located above and on one side of the web; it comprises an elaborate chamber and corridor into which food remnants frequently are woven. The retreat is connected with the catching web through *2.* the *bridge*, 10—15 cm long, and consisting of two coarse-meshed layers. *3.* The *catching web*, has a horizontal platform, fine meshed, with a circular center surrounded by short radii; about 20 vertical threads, around 20 cm long, connect the platform with the ground; the vertical catching threads are viscid in their lower 2—5 cm. Complete construction of such a web takes many days, and after that it is continuously further elaborated over a long period of time.

Differences in the two web types are apparent in construction time (30 minutes versus several days), geographical distribution of components (all parts interwoven versus separation into three distinct subunits), and regularity (relatively regular, nearly rectangular meshes versus some coarse hexagonal to octagonal reticulum and much irregular structure). Similarities between the web types can be seen in the use of different thread for separate functions (scaffolding versus catching parts with appearance of glue on the latter), in the relatively high position of the retreat (which permits quick descent to the prey), and in some of the building and fastening processes. But it appears to this writer that similarities are more a matter of similar purposes and use of similar material, than based on a common origin of web patterns in the history of spiders. I see no good reason for the assumption that any one web type should be the predecessor of any other.

IV. Altered Web Patterns

Regarding webs as a record of spiders' movements during a short period of time, and translating web measurements into movement patterns, a basis exists for an application of the spider-web method to drug studies. In order to understand the advantages as well as the limitations of this application of web-building, some of the methods used in drug studies and other behavioral investigations will now be discussed in detail. This should serve the additional purpose of enabling other investigators to repeat and amplify our studies.

A. Methods of Investigation
Procurement of Webs in the Laboratory

Whether we deal with the geometry of an orb web built by an undisturbed animal, or whether we interfere with building in specific ways in order to evaluate the consequences, the method of recording and data processing is basically the same. We make the webs visible by spraying them with white paint, photograph the sprayed webs, and measure and evaluate the patterns in terms of plane geometry. Recently better lighting conditions and more sensitive photographic film has permitted good photography of unsprayed webs. Most experimental effects are sufficiently described as departures from the geometric pattern of the normal web, but because so many factors influence daily web-building, many webs have to be evaluated in order to obtain statistically valid data.

The experimentally desirable achievement of obtaining one web per spider every day all year round in the laboratory is fore-stalled by a number of problems. Inadequacy of control sometimes becomes apparent only belatedly, to cast doubt on findings previously attributed to the experimental variable. For example, systematic difference in webs built in the laboratory on Mondays eluded detection for some time before it became evident that, unless a web was destroyed by the end of the day, it was probable that the spider would eat it and as a consequence build a larger web on the next day (BREED et al., 1964). This discovery led to the simple procedure of not using Monday webs as controls in experiments. Other problems are less amenable to solution.

Some web changes are associated with maturity of the spiders (WITT and BAUM, 1960). The simple solution would seem to do experiments with mature animals only. However, there is really no "mature" web: after the female has reached full size, other factors, such as egg development, affect weight and hence the web pattern; and males build too infrequently in later life to be experimentally useful. Body weight varies markedly even with spiderlings of the same cocoon; therefore, age is not a reliable measure of animal size. The range of weights for one set of experiments e.g. has varied from 6 to 800 mg; in this group the median weight was 118 mg (REED and WITT, 1968). No attempt was made therefore to restrict the size of the animals used in the experiments; the control webs were those built by each spider immediately previous to the drug-administration, and drug dose was adjusted to individual body weight.

During winter months, the mortality rate of spiders is high, and weaving is relatively infrequent, even in a warm laboratory. The low frequency of building can be attributed in part to the relatively even temperature in the laboratory and to

Table 7. *Parameters of webs built by light and heavy spiders. All came from one cocoon, but were given different amounts of food*

No. of animals	Body weight mg	Thread length m	No. of radii	No. of spirals	Spiral area cm²
20	19.6 + 11.3[a]	11.38 + 4.78[a]	31.9 + 5.5	31.4 + 7.9	16.5 + 8.5[a]
20	138.7 + 35.9[a]	16.23 + 4.8[a]	27.3 + 3.8	32.4 + 6.7	35.3 + 11.2[a]

[a] Significant difference below the 1% probability level between corresponding groups.

brevity of daylight (WITT, 1956). The matter has been improved somewhat by installing a programmer which controls temperature and humidity, providing a temperature minimum in the early morning hours, followed by a steep rise, and by installing artificial illumination which simulates 16 hours of daylight. In spite of extensive experimentation with temperature, light, humidity, and other environmental variables (see SPRONK, 1935; WOLFF and HEMPEL, 1951; WITT, 1956), it appears reasonable to state that we do not know and are unable to reproduce in the laboratory the combination of circumstances which is responsible for a certain web-building frequency, nor for the timing or "release" of the process. Among a multiplicity of factors, the following have been found to have some influence: the animal's inner rhythm, a change from dark to light, a steep rise in temperature following a temperature minimum, a full silk supply, hunger, weather conditions, barometric pressure, the presence of flying prey.

Egg cocoons can be hatched all year round but, for reasons not yet clear, only relatively few cocoons release spiderlings in the laboratory. About 200 animals can be obtained from one cocoon. Except for the tedious job of feeding all spiderlings, raising them on a diet of fruit flies and water presents no difficulties. It is interesting, although experimentally inconvenient, that even "litter-mates" from one cocoon, raised under similar conditions, grow at different rates. Table 7 gives an example.

Frequency of web building varies with species: female *Zygiella-x-notata* Cl. and *Araneus diadematus* Cl. build daily, with interruptions during molting; the former can be easily caught out of doors in central Europe, the latter in Europe and the northern part of the United States. While *Zygiella* lives for only about 8 months, *Araneus diadematus* can survive for 18 months. *Araneus sericatus* Cl., although convenient in that it can be caught out of doors nearly all year round and in that it builds webs very similar to those of *Araneus diadematus*, spins less frequently, about every other day.

Obtaining the Record

Adult female cross spiders *(Araneus diadematus* Cl.) are kept in the laboratory under controlled temperature and light conditions in individual $50 \times 50 \times 9$ cm aluminum frames with two glass doors and screen sides, built by State Metal Awning Co., 131 Shonard Street, Syracuse, N.Y. The animals receive water daily and house flies twice every week. Experience has shown (WITT, 1956, 1963) that under these conditions an orb web is built nearly daily during 20 to 30 minutes around 5:30 a.m.; at this time the light is turned on in the laboratory and the temperature rises steeply.

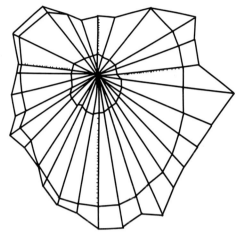

Fig. 18. The web as the computer "sees" it: in terms of the position of a set of intersections of threads. Rather than measuring all such crossings (more than 1000 in the usual web) points have been reduced to those marking angular position of radial threads, those locating innermost, outermost spiral and frame thread on each radius, and all spiral positions in each of the 4 cardinal directions

The daily built webs are fragile and hardly visible. For behavioral analysis moving pictures are problematic because of the low photographic contrast of the thread, to say nothing of the expense and labor of accumulating and analyzing films. Homann's method (See PETERS et al., 1950) of placing cases with complete webs into a box filled with smoke of ammonium chloride was an early solution to the problem of visibility of the orb; we find it easiest to remove the spider and coat the threads lightly with Krylon white glossy spray paint.

The web is photographed with 35 mm high-contrast film. So that each thread will be reproduced as a thin white line contrasting sharply with the background, the case is placed in front of a black box and lighted from four sides. A weight, composed of a lead bar and three cotton threads 20 mm apart, is suspended in the plane of the web to calibrate for enlargement and for direction of gravity (Fig. 1).

The record is thus obtained quickly and stored readily. The experimenter may return to it in order to measure selected aspects of the pattern: data on drug effects were gathered in this manner.

Measurements. In order to convert the web into numerical form representative of the pattern, two forms of measurements are made:

The position of each radius is measured in degrees starting at left horizontal (0° and 360°). The radii do not converge toward a single point but a locus in the plexus of threads forming the hub; the initial selection of the center is therefore a somewhat arbitrary one. Moreover, the radii, although straight during their construction, subsequently are distorted by the spiral thread. Measurements of radial position therefore are taken at the intersection of the radius with the frame thread.

Having established the compass position of the radius, three points are measured on it: the distance from the center to the first spiral thread, the distance from the center to the outermost spiral thread, and the distance from the center to the nearest frame thread. All measurements are in mm. The non-viscid thread forming the hub is not measured.

In addition to these four measures on each radius, the distance from each spiral turn from the center is measured along radii next to the four cardinal compass direction. In parameters derived from the raw measurements, the cardinal compass points are used as representative of a quadrant of the web.

Limitations on accuracy of measurement are imposed by the present techniques. The film is projected to full size and, depending on the quality of the photograph, threads appear more or less distinct. A ruler, pivoted on a compass is centered on the web and readings taken where the projected image intersects the ruler and compass. Readings are made to the nearest degree or millimeter.

The dimensions of the web, as received on punch cards by the computer, are shown in an example in Fig. 18.

Computations

The computer program has been changed as results of experiments suggested possibilities for improvement. The first program is described in some detail by REED et al., 1965. At present the following measures are generated by the computer[1]:

1. Number of radii: all the radii are counted, total number listed.

2. Oversized angles: definition: an angle is to be called oversized when twice or more the size of the smaller of its neighbor angles; the number is listed.

3. Median angle: all angles are arranged according to size and the one is selected which is half-way between the largest and smallest.

4. Angle regularity: in the normal web, each central angle differs but slightly from its neighbors, although upper (North) sectors tend to have larger angles than do lower (South) sectors. The mean discrepancy by which all adjacent angles in one web differ from each other is taken as an index of regularity of placement of radii.

5. Standard deviation of central angles: we use the familiar formula for standard deviation:

$$\text{SD} = \frac{1}{N-1} \sqrt{N \Sigma x^2 - (\Sigma x)^2}$$

where N = number of angles

x = central angle size in degrees

---

[1] The authors will be glad to provide the whole program on request.

6. Width over length: by dividing the horizontal diameter of the catching area by the vertical diameter, a figure for the oval shape of the web is obtained.

7. Radius North over South: the vertical position of the hub with respect to the perimeter of the spiral zone is established by dividing the distance from hub to upper limit of spiral by hub to lower limit.

8. Radius East over West: this is a measure of the horizontal position of the hub in the spiral-covered web area, and an expression of the right-left symmetry of the catching area.

9. Center area in mm square: computation of the central area is accomplished by summing the areas of the sectors formed by each central angle and the first spiral thread. Note that the central zone includes the densely-woven hub and the spiral-free zone.

10. Spiral area in mm square: the spiral area is computed by subtracting the central zone from the summated area formed by the central angles and the last spiral turn.

11. Frame area in mm square: by frame area we mean the space lying between the outermost spiral turn and that irregular boundary which is formed by the peripheral end of the radii (where the radii are moored to frame threads). Because of the irregularity of that outer boundary, a simplifying and not entirely satisfactory assumption is made regarding the frame area. A chord closing the sectors at the distance of the radius-frame intersection is assumed, and the areas of the spiral and center zones subtracted from the summated sectors formed by the central angles and the assumed chord. This procedure yields a frame more regular than the true frame and underestimates the frame area.

12. Frame over spiral area: the quotient of the two areas gives a measure of the catching or functional zone of the web relative to the above defined frame area.

13. Center over spiral area: the ratio of these two areas indicates the relative amount of the center area of the web which has been left nonfunctional by early or late termination of the process of building the spiral from the outside in.

14.—17. Number of spirals West, North, East, South: the number and location of spiral threads are read into the computer for four cardinal directions. In the normal web, the number of spirals is not uniform in every direction. A comparatively large number of spiral threads is placed in the lower portion of the web, the spider reversing its direction while building additional threads. Webs are built toward one of the upper corners of the square frame; as a measurement convention, the corner at which the web is built is designated West.

18. Thread length: the total thread length, viscid and non-viscid, is computed as follows:

a) the lengths of all radii are summed,

b) at the frame distance, an arc closing the sector is assumed and added to the radial sum,

c) the distributions of spirals in the four cardinal directions are taken to represent four quadrants. The arcs within each quadrant and at the distance of each spiral turn are summed. This is a simplifying assumption; the actual spiral forms many small chords. The result is added to (a) and (b).

19.—22. Relative deviation of spiral turns West, North, East, South: although there is a gradient from center to periphery, the turns of adjacent spirals are rather evenly spaced. Using the difference between adjacent spacings, a measure was derived which subsequent analysis showed to be contaminated by size (WITT and REED, 1965). The present formula takes size into account:

$$RD = \frac{SD}{\bar{X}} \quad \text{Where SD} = \frac{1}{N-1}\sqrt{N \Sigma X^2 - (\Sigma X)^2}$$

where N = number of spiral turns in any of the four directions.

$$X = \text{first difference between entries} \quad \bar{X} = \frac{\text{sum of first differences}}{\text{(number of entries)} - 1}$$

23. Mesh width: the functional snare formed by the spiral zone is dependent upon the separation of threads occupying that area. The number of radii is inversely related to the mean central angle, and is taken as a measure of the average angle. The following formula is employed:

$$\text{Mesh width} = \frac{\text{spiral area}}{\text{(Number of radii)} \cdot \text{(Mean number of spirals)}}$$

This formula expresses distribution or density of the threads in the spiral area.

24.—25. Median mesh size sample North and South: in addition to using the ideal mesh as a measure of fine structure, a sample of meshes is taken in 3 North (clockwise from 90°) and 3 South sectors (clockwise from 270°). This measure, in contrast to the ideal mesh size, depends upon the distribution of central angles and spiral distances. The median is calculated.

26.—27. Standard error of mesh size sample North and South. The standard error indicates an irregular web's departure from normal distribution of meshes.

$$SE_{med} = \frac{(N-1)^2 \, 1.253}{\sqrt{N}} \sqrt{N \Sigma X^2 - (\Sigma X)^2}$$

Current evaluation of the photographs consists essentially of three steps: first, measurements are made of selected points (see above); then components of the pattern are calculated from the figures (see above); and finally statistical comparisons are performed. Each of the 27 parameters is tested for differences in a T-test of the control and the experimental webs of each treatment. The Chi-square analysis is also performed for the measures concerned with regularity of thread placement, because of the categorical nature of the measure. We set significance at the 1% level of confidence for all tests. In such a large number of tests, spurious significance effects are to be expected with a frequency set by the significance level; replicated analyses assisted in evaluating such circumstances.

In the usual drug experiment, a great number of statistical tests is performed (27 tests per dose-level and time-period). This enhances the probability of obtaining significant differences. To compensate for this circumstance, we have chosen to make errors of type 2, i.e., to miss real differences, by setting a somewhat stringent level of significance. We have also examined the results of two sets of summer experiments for consistency of significant differences for all measures (REED and WITT, 1968). Not all experiments were repeated in both summers and modifications were made in the second series on the basis of experience in the first. It was found, for instance, that the first series extended too far into late summer when the animals were beginning to show seasonal decline in frequency of web building. While comparison is limited for these reasons, there are 205 pairs of T-tests in which it is justifiable; 21 of these reached the 0.01 level of significance one summer but not in the other. The direction of change was consistent; however, pooling the data for two summers produced 0.01 level significance in 20 of the pairs. The experimental differences found for spiral number, thread length, angle regularity, radial number and spiral zone are particularly consistent.

Drug Application

Some spiders were given drugs more than once, though never more than 3 times. At least 2 weeks intervened between drug-administrations to the same animal.

Under laboratory conditions, all spiders building a web on a given morning do so within a few minutes of each other. Drug administration was timed to precede the building-period by 12 and 24 hours. An interval of 24 hours must lapse before another web can be expected, hence the course of the drug effect in experiments is measured conveniently at 12 hour intervals (i.e. up to 48 hours in diazepam experiments).

The spiders preoral digestion and its ordinary drinking habits make oral administration (drug in sugar water) a straightforward procedure (Fig. 19). Each animal is weighed and the volume of drug solution (or suspension) of known concentration adjusted, with the help of a Hamilton microsyringe, so that an animal receives the desired dose with an accuracy of better than 10%. The spider is held while it drinks in order to insure digestion of all the fluids. However, it is possible that the animals later regurgitate some of the solution, particularly that of high concentrations with strong taste.

If injections are performed, it is important to keep the volume as well as the injection needle as small as possible. The Hamilton microsyringe again has proven convenient. In repeated experiments no difference in the quality of effects was found between the two ways of administration (CHRISTIANSEN et al., 1962).

Apart from the web measures, it may also be desirable to evaluate the spider's use of the material available in the ampullate glands; these glands produce all the scaffolding for the web (PEAKALL, 1964). A sample of webs which appear to be smaller or sparser (as a result of a drug i.e.) can be digested; nitrogen content is determined in the digest in order to establish the amount of material expended in construction.

For this purpose the catching area is cut free of the web frame. As radii are severed just outside the outer spiral thread, the web collapses into itself so that finally it appears as one line attached at both ends to the last intact radius. One end is then attached to the top of the frame with a piece of tape and the other end cut free. The web is folded upward with the free end onto low nitrogen filter paper which has been cut into small discs weighing between 3 and 16 mg. The discs are weighed, the web added and the discs plus the web reweighed on a Cahn electro-

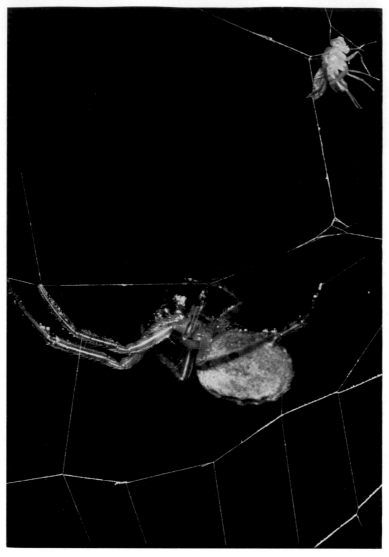

Fig. 19. An adult *Zygiella-x-notata* is enticed from its retreat by a Drosophila in the web and a drop of liquid is quickly applied to its mouth parts. No longer concerned with the fly, it hangs quietly in the web ingesting the drop — which may contain a drug in solution. (Photographed by J. J. WEITMANN)

magnetic balance sensitive to one microgram. It can then be placed in the test tube in preparation for nitrogen determination (see page 17).

B. Changes of Webs with Age

There can be no doubt that a small baby spider builds a web different from that of a full-sized animal. Several questions can be asked as to the changes: Is there a continuous change in web size and pattern from babyhood to adulthood and old age, or are there discontinuous changes, certain periods in which distinct parameters are altered? We can also ask whether the pattern changes are related to the growth of the body of the spider. In any case, if we want to compare patterns of different species, if we want to compare webs of different individuals, if we want to establish long range effects of drugs or lesions, we have to know something of the changes of webs as the spider grows and ages.

Several authors have published results from their observations on webs of young and adult spiders of different species. These results are contradictory. WIEHLE (1927) counted more radii in the webs of young spiders than in those of adults, while KOENIG (1951) found fewer radii in the smaller webs of young *Araneus diadematus* than in webs of full-grown individuals. SAVORY (1952) observed a slight decrease in the number of radii during four summer months in outdoor *Araneus diadematus* webs, but points out that there was considerable variation. TILQUIN (1942) — who counted radii in many webs of *Argiope bruennichi* which he kept indoors — related number of radii to molting. He describes the slight increase in average number of radii (from 22 to 29) around the third molting. For the rest of the spider's life the average number remained relatively constant. His counts varied between 10 and 40 radii in the webs of young spiders as compared to between 19 and 41 in adult webs. PETERS (1953) measured mesh size between two radii and two turns of the provisional spiral, in young and adult *Araneus diadematus* webs. He found that there was a change in the ratio of body size/mesh size during the life time of a spider. This led him to the conclusion that body size does not determine mesh size.

WITT and BAUM (1960) raised *Araneus diadematus* Cl. from an egg cocoon in the laboratory. Spiders were left together in a glass jar until they began to attack each other. At that time they were put into individual jars and supplied with three and one-half by five cm cardboard frames. Eleven days after the first spiderling had left the egg cocoon, several webs were photographed; they represent either the first or very early webs of each spiderling.

Similarly the first webs of *Neoscona vertebrata* McCook appeared 2 weeks after the cocoons had been placed in glass jars. Again all spiders were kept in frames of increasing size and fed amply. The light and temperature in the laboratory was controlled as described above.

There was no uniform or steady increase in the web size during the life time of *Araneus diadematus* Cl. A fast increase in the catching area (67%) from June to July was followed by a slower increase (17%) from July to August. For the next 3 months the size of the catching area did not change significantly.

The numbers of radii and spiral turns were correlated, but neither was correlated with web size development. A maximum number of radii and spiral turns was observed in webs built in July when they were yet to attain full size. Compared to the size of the webs, the 2 month old spiders had a small mesh size. With growth of the spider, the meshes became increasingly large. These changes in webs of growing *Araneus diadematus* can be summarized: during the first month of life the spiders built small, closely knit webs; in the second month larger webs with small mesh size were observed; large webs with increasing meshes were characteristic for older spiders.

The pattern of changes in webs of growing *Neoscona vertebrata* was different from that of *Araneus diadematus*. The web attains its full compliment of radii in January, following an initial decrease. During this time, the catching zone remains fairly constant in area, but in March it shows very rapid increase, to a size more than twice the area at the beginning of the growing spurt. (The final catching area is 2½ times that of an adult *Araneus* web). Since radial and spiral turns show no similar increments, mesh size increases with age.

It is interesting to compare the changes in web pattern with those in body weight and leg length. Leg length for *Araneus diadematus* increases rapidly (47%) during the first month and more slowly later (about 10% per month in four months). Body weight increases more uniformly. This manner of growth accounts for the relatively light spiders with long legs which are found outdoors in July and the heavy long legged spiders found in November.

The average body weight of *Neoscona vertebrata* in June (the last month of their lives) was 128 mg. The average leg lengths of three individuals for the same month was 20 mm. In comparing these figures with those of full grown *Araneus diadematus*, we find that the older *Neoscona* were considerably lighter and possessed significantly longer legs. No measurements of younger *Neoscona* have been obtained.

The patterns of growth have no apparent relationship to ratio of length to width of web (vertical to horizontal diameter), to the symmetric position of the hub in the catching area (East over West), or to the regularity of angles between radii.

Bodily dimensions determine some web parameters, but the variable relationships between body weight, leg length and web size require cautious extrapolations from webs built at one age to those built earlier and later. Failure to note this variability may explain some of the contradictory statements in the literature. Our data confirm PETERS' observation that there is no simple relationship between leg length and mesh size in *Araneus diadematus*. There was, in our webs, a relationship between leg length and web size and another similar relationship between body weight and mesh size. The young *Araneus diadematus* with relatively long legs and a light body built large webs with narrow meshes. When the spiders became heavier without showing a comparative increase in leg length, their webs showed increasingly larger meshes. A similar relationship holds true for webs of heavy and relatively short legged fullgrown *Araneus* as well as those of the light, long-legged adult *Neoscona*. The webs of the latter were much larger in size and contained more spiral turns than those of *Araneus*.

These statistics, however, smooth out day to day changes in web size and measures. As an example, shortrange change in the webs of a single spider was measured on August 13, August 24, and August 31. The size increased from 542.4 on the first day to 770.8 on the second day and decreased to 229.2 cm^2 on the third day. This represents 67% decrease in catching area in one week in webs of one spider. In this case body weight and leg length were apparently not changed; we conclude that these two body dimensions are not the only factors determining the web pattern. Their usefulness is limited in the prediction of one web for a certain day, but they permit general statements for a sample of webs[1].

---

[1] Another change is the appearance of the hatched band in the web. This can appear earlier or later, it can consist of one hatched band in the early stage, and two in the later stage. Again as far as we know there is no correlation found with any known parameter and the appearance of these web characteristics. It is even observed that a spider which has already built a hatched band, can build a web without that later.

The influence of body weight on web characteristics can be demonstrated by attaching weights to spiders at certain times of life. This is the subject of the discussion in the next section.

C. Changes of Web Pattern with Changes in Body Weight

MAYER (1953) was, as far as we know, the first to attach a weight to the body of *Araneus diadematus* Cl. and compare webs built before and after this procedure. In the course of 34 days of the experiment, her one animal increased in body weight from 21 to 80 mg through natural growth, and the attached weights varied from about 100 to 175 mg. In every one of the four experiments the spider built a web with fewer radii, wider-spaced spirals, and a slightly increased eccentricity of the inner spiral. MAYER does not try to interpret the changes, except for attributing the slight eccentricity to increased effect of gravity.

In our experiments (CHRISTIANSEN et al., 1962) we tested the working hypothesis that a heavier spider has to build a thicker thread in order to hold its own weight. If the thread becomes thicker, with a limited amount of material being available, it necessarily becomes shorter. Therefore the body weight of spiders was increased by about 30%; thread length, web weight, and web pattern were compared before and after the addition of weights.

Increasing weight of spiders. Eight *Araneus diadematus* and five *Araneus sericatus* were used in the weight experiments. Two of the spiders were used twice. As control, eight *Araneus diadematus* and six *Araneus sericatus* had only adherent applied. Each spider was weighed the morning after its web was built. A piece of lead equal to 30% of the spider's weight was placed on the dorsal surface of the abdomen while the spider was held down with a forceps. Ace adherent was used under the lead piece and collodion dissolved in an ether-alcohol mixture over it, sealing the edges. The spider was immobilized until the glue had dried. After the next web had been built the following night, the weight or adherent was removed by carefully cutting under and around it with small scissors. In addition, each animal served as its own control.

Measuring thread length. The glass doors were removed from the spider box and a piece of black cardboard placed behind the box to make the web more easily visible. The number of radii and spiral turns was counted and measurements were taken of the NS and EW diameters both of the catching area, determined by the outmost spiral turn, and of the free zone in the center, bounded by the innermost spiral. The following formula was developed for us for an approximate calculation of total thread length of the catching area of a web, assuming that the spiral consisted of even-spaced perfect circles:

$$L = n \cdot R + k\pi (a + R)$$

In which L equals thread length, n equals number of radii, R equals radius of a circular catching area $(N + S + E + W)/4$, k equals number of spiral turns counted along S radius, a equals radius of inner circular area (free zone).

Thread length. In 15 webs built after the body weight had been increased by 30%, the total thread length decreased significantly from an average of 15.95 m $\pm$ 1.69 to an average of 10.76 m $\pm$ 0.95. Control webs of spiders with adhesive showed no similar effect. The first day after removal of the weight 14 spiders built webs

with significantly shorter thread of an average of 9.86 ± 0.85. 3 days after the weight
had been detached, thread length increased to an average of 12.25 m ± 0.95: no
longer different from initial control.

Thread thickness. According to the hypothesis the heavyer spider builds a
thicker thread which, because of the limit in quantity of material, has to be shorter.
In 13 control webs an average of 11.5 ± 2.1 µg protein/m thread was measured.
After addition of 30% weight, the ratio had significantly increased to 22.6 ± 5.4 µg/m.
Again, on the day after weight removal, thread remained thicker with 21.3 ± 7.1 µg/m
in ten webs, and returned to control levels with 14.1 ± 1.5 3 days later. The adhesive did not change thread thickness significantly.

All evidence is in favor of the hypothesis that the spider makes a heavier thread,
in order to support its increased weight. It seems that it does not have enough
material to build a full-length thread. Removal of weight does not have the immediate
effect which addition shows: the web returns slowly to control values. This could
be explained by some damage done through cutting the weight off, or by assuming
a kind of "learning" process which is not immediately unlearned. The adjustment
of the whole web pattern to the new condition of "increased weight — shorter
thread" points toward a central mechanism of information processing which is more
than a reflex. This point will be discussed more extensively later.

D. Dependence of Web Geometry on the Number of Legs

REED et al. (1965) measured webs of *Araneus diadematus* before and after elimination of one front leg or two front legs. The purpose of the experiment was to
establish statistically any change immediately following elimination of the leg, change
over time, possible compensatory adaptation, and the difference in individual reactions
between spiders. For the first time the computer was used in experiments analyzing
the great number of data embodied in the web. The extensive survey of thread
position so provided was directed in this case to elucidating the locomotor as well
as the measuring function of the front legs.

Leg-performance experiments had preceded these, usually performed with one
animal. JACOBI-KLEEMANN (1953) in a detailed photographic analysis of the locomotor function of each leg of *Araneus diadematus* Cl., observed a groping of the
first leg during spiral placement. The movement was distinguishable from locomotive
function. The second ipsilateral leg appeared to assume this palpating function in
the absence of the first leg. She made a frame-by-frame film analysis of web building
spiders deprived of one or two legs and described only one finished web.

HINGSTON, as reported by SAVORY (1952), produced a displacement in the relative
positions of spirals, apparently as a consequence of providing the first leg with
erroneous placement cues. To test this inference, the tip of the foreleg was removed; in
SAVORY's words the result was "a badly made, untidy web". (See also LEGUELTE, 1965a).

In the first experiments of REED et al. (1965) legs were cut off with scissors near
the trochanter. It was observed that under those conditions some body fluid leaked
out. In order to keep the trauma as light as possible, the autotomy mechanism which
exists in spiders was used in later experiments: the animals were carefully held between two fingers and a leg was pulled off with the help of a forceps. Like JACOBI-
KLEEMANN no difference in the consequence of both procedures was observed.

Of the 15 spiders deprived of one first leg (-1 spiders), 11 built a sufficient number of webs to warrant statistical analysis; four of the six spiders deprived of two first ipsilateral legs (-2 spiders) built a sufficient number of webs. The first webs were constructed within the week following leg removal, usually the next day, with the exception of one spider who built her first web after 11 days. Altogether 85 control and 102 treatment webs were analyzed.

Table 8

Variable	Control		Experimental		Effect of leg absence on	
	$\bar{x}$	sd	$\bar{x}$	sd	early webs	late webs
Angle regularity	4.9°	1.36	7.84°	2.70	less regular	less regular
Number of radii	23.13	4.08	18.40	3.09	fewer radii	fewer radii
Median angle	15.30°	3.13	18.83°	2.72	larger angle	larger angle
Number of spiral turns						
West	18.13	6.40	13.67	8.93	no change	no change
North	17.80	6.35	12.80	8.26	fewer turns	fewer turns
East	23.07	7.82	16.60	7.86	fewer turns	fewer turns
South	27.00	7.41	19.06	10.00	fewer turns	fewer turns
Spiral regularity (mm)	1.43	0.30	2.01	0.73	no change	no change
Frame area (cm²)	148.36	43.25	132.36	72.68	borderline reduction	
Spiral area (cm²)	327.34	115.22	216.84	108.13	borderline reduction	
Center area (cm²)	30.08	8.17	27.58	8.39	borderline reduction	
Ratio:						
frame/spiral	0.491	0.185	0.500	0.277	relative reduction of spiral	
center/spiral	0.098	0.040	0.141	0.082	relative reduction of spiral	
N/S	0.670	0.138	0.719	0.068	no change	no change
E/W	0.828	0.114	0.838	0.068	no change	no change
Width/length	0.848	0.098	0.957	0.176	no change	no change
Thread length (m)	13.38	5.11	8.55	5.13	red. length	red. length
Mesh width (mm²)	66.44	12.23	80.92	26.53	no change	no change

From each of the 15 spiders (with one or two legs missing) one control and one experimental web was chosen randomly and means and standard deviations are listed. To test the onset and permanence of the effects, all webs built just prior to delegging were compared with all webs built immediately after, and with webs built at least 3 weeks later. Entries in this case mean that the change was significant at the 0.01 level of confidence.

Results are shown in Table 8.

Removal of the first or first plus second leg apparently results in the production of webs with significantly fewer radii, less regularly spaced than in controls. Radii were reduced in number by slightly less than 20% in the average; the median angle was by consequence increased significantly. In considering the data spider by spider, i.e., by testing each spider's total normal production against its total production when minus a leg, four of the minus one spiders failed to show the radial and central-angle effects at the criterion level. While this result may in part be a reflection of the power of the T-test with small numbers, it seems also the case that either variability in web pattern is sufficient to mask effects for the individual spider or that consequences of loss of one of the first legs are not identical for all spiders.

Reduction in the number of spirals is slight but significant with the exception of the west quadrant, which is reduced, but not to criterion. The effect is observable in the first webs after delegging and it is permanent. Absence of a single leg seems sufficient to reduce the number of spirals built.

Results in respect to spiral regularity, i.e. the even spacing of the spiral threads, are less clear. As in the case of the measure for differences between adjacent angles, distances were used to express regularity. The mean difference for all four quadrants is shown in the table. Comparison of the last control webs with the early and late experimental webs does not indicate a reduction in regularity of the spirals. However, the data and tests reported in the Table 8 are derived from mean scores. The distribution of the differences is narrow and skewed (as in angle differences); hence separate distributions for all control and all experimental webs for each spider were compared by means of the Chi-square test. Disturbance at the criterion level was found for all of the -2 spiders and for 5 of the 11 -1 spiders.

Since the Chi-square analysis is performed on all control and all treatment webs of each spider, the temporal course of effects cannot be gauged. When distributions are prepared for final control webs of all spiders to be compared with final experimental webs, a significant Chi-square is not obtained. This finding suggests that spiral regularity may be reinstituted through use of alternative methods of spiral construction. But, for some spiders at least, irregular placement of spirals is a lasting consequence of the absence of the first or the first and second legs.

If the probing function of the first outer leg is of decisive importance in spiral construction, the spider could compensate for the loss of one front leg by using the intact leg on the outside. This procedure would result in fewer pendulum turns in the spiral, and relatively more circular construction would take place. Such behavior was predicted by JACOBI-KLEEMANN, but not confirmed by our data. The relative number of spiral turns with respect to direction is unchanged in the treatment webs. The data suggests that the -1 spider as well as the -2 spider rather builds an irregular spiral than changes its procedure of turning back and forth during spiral construction.

The spiral area contains the viscid threads and hence constitutes the functional zone of the web. It takes the longest time to build and requires probably a greater amount of body movement than any other part. Its reduction may signify a sparing of effort in the 7 and 6 legged spider. As is evident from the standard deviation in the Table 8, the spiral area varies widely in magnitude.

The findings of the relative position of center and shape of web may be summarized as indicating that the oval shape as well as the position of the hub with respect to the perimeter of the spiral zone stay unchanged. Webs built under the influence of drugs which alter central nervous system function can show severe change in the position of the center and the shape of the web. The contrast between that finding and the constancy measured after delegging points to a central regulation of web shape and hub position, delegging contributing only a peripheral treatment.

Thread length varies greatly even in the normal web, as is evident in the standard deviation in the Table 8. The longest total length in the normal sample, measured by the above procedure, was 34.6 m, the shortest 6.0 m.

Reduction in total thread length is most clearly evident in the test performed in the last control versus the early and late experimental webs. The effect is immediate and permanent. Every spider's mean thread length was lower by at least 13% in the experimental condition. There is some sign that the effect is greater for the -2 spiders, but the small numbers do not permit a clear conclusion. The mean reduction in thread length for the -1 spiders was 40%; for the -2 spiders 58%. If we take thread length also to indicate the length of the route taken by the spider during web-building, this change would imply the same tendency as does the relative decrease in spiral area: decreased total locomotion.

Mesh size is interesting because earlier experimental results (WITT, 1963) indicate that if the spider intends to cover the same area with less thread (if less total length is available), it resorts to a wider-meshed web. The smaller area with unchanged mesh size in our experimental webs again points to a saving of labor rather than to shortage of thread material.

From these experiments it is concluded that the first leg has a measuring as well as a locomotor function and that the second leg can partly substitute for the measuring function. Web changes following delegging reflect a loss of normal adequacy in both of these functions. No evidence of right-left differences in leg function were found. All adaptation to the absence of one or two legs had occurred in the first web after the operation.

E. Feeding

There is general agreement in the literature that heavy feeding is followed by several days without web building (KOENIG, 1951; WOLFF and HEMPEL, 1951; WIEHLE, 1927; PETERS, 1932) the interpretation of such behavior usually runs as follows: the spider is no longer hungry and consequently does not need a trap for catching food. Expressing it in a different way, we may assume that the hunger drive is too low for releasers like temperature and light to operate. Satiation somehow inhibits web-building. As far as we know no data bearing on this idea had been collected through experiment.

The effects of food deprivation or hunger are more complex than the effects of satiation. Two trends may work in opposite directions: Frequent web-building and larger webs provide food; infrequent building preserves the highly organized protein which composes the precious web-material. It has been observed many times that a spider collects and eats its old web (BREED et al., 1964). Failure to observe this phenomenon may be responsible for contradictions in the literature regarding web-size during hunger experiments.

In our experiments (WITT, 1963) three parameters were observed under different conditions of feeding: the size of the orb webs as measured on photographs, the amount of thread produced in the glands as established from thread reeled out onto the axle of the motor, and the body weight of the spiders. It was found that by feeding a 12.5 mg fly to a spider with a body weight of 138 mg 5 days each week for 2 weeks the body weight increased to 159 mg. An average 0.3 mg of thread per day could be pulled. Though figures for thread production varied widely from day to day, no systematic decrease was observed. We concluded that the spider showed normal growth and balanced silk production. The average web size of a population of spiders during that period of time increased in young animals and did not change in adults (compare with "growing webs", WITT and BAUM, 1960).

The webs of 15 adult spiders and 13 small, young spiders (8 to 10 mg body weight) were compared with the webs they built after 1 and 2 days of food deprivation and after daily elimination of webs. No statistically significant changes in web size or web proportion (number of radii, number of spiral turns, diameter) were observed, nor was there measurable change in the amount of thread pulled or in the body weight of the animals.

A group of 9 spiders had their webs eliminated daily and were kept without food for 6 days. Their average web size showed no significant decrease after that time, but less thread could be pulled. It dropped from an average of 0.49 mg to 0.39 mg daily per animal, and body weight decreased by 12%.

After 10 days of food deprivation the same group of spiders on the average built significantly smaller webs of unchanged structure: thread production had further decreased to 0.30 mg per day per animal, and they had lost an average of 49% in body weight.

Another 17 spiders were kept without food for 20 days. After that period their webs had become significantly smaller and wider-meshed: they were built with less thread. The body weight of the animals as well as the thread which could be pulled had decreased, but we could still obtain daily 47% of the thread weight which they had produced before deprivation.

After feeding had been resumed, no change in web size or body weight could be observed during the first 3 days. But 10 days' feeding resulted in a significant increase in the average body weight without changes in web size: the webs stayed small.

In interpreting these preliminary results it appears of interest to note the sequence in which food deprivation and refeeding affected the different parameters. During the first 6 days of hunger, thread production and body weight had decreased; large webs still were built but with obviously thinner thread: even when the animals could no longer produce the full amount of silk, large webs sufficient to insure a good catch of prey were built. In later periods of food deprivation, webs became smaller and wider meshed, reflecting the severe decrease in thread synthesis. But some silk production still went on, obviously at the expense of other body constituents, as shown by the surprisingly high loss in body weight. It appears that the production of silk stands so high in the hierarchy of vital functions in spiders that it is retained while other tissues shrink. On resumption of feeding, however, thread production and web area do not increase immediately. The increase in body weight with consistently small webs indicates that the spiders first rebuild their lost tissues. It is also of interest to note in the results of this experiment the close coordination of behavioral (web-size) and biochemical (amount of thread) regulatory mechanisms in an emergency situation.

In addition to quantity of food, the influence of its quality on web-structure and silk quantity has been tested (BREED et al., 1964). In our ordinary laboratory procedures, the web of a spider is eliminated every day to permit construction of a new one without use of possible remnants of the old. However, preliminary observations indicated that new webs, built when the old ones had been left intact, showed significant changes.

Direct observation of *Araneus diadematus* Cl. throughout the night as well as scanning of spiders which sat on radioactively labeled webs shows that the majority of animals *in the laboratory* systematically destroyed their old webs, sector by sector, in the course of a night. They collected the material and digested it, building new webs in the early morning hours. These experiments permit no conclusions in regard to the behavior of outdoor spiders. Spiders outdoors may renew with greater or lesser frequency. In the laboratory the old web material, under such circumstances, served as food, and the intact structure of the old web may have been a signal for the animal, influencing its behavior toward new construction. To distinguish between these two possible cues, one group of spiders was left with its old web intact, while a control group was left with the old material assembled in an unstructured ball. The results are shown in Table 9.

The table shows that in contrast to the effect of the wad of material, intact webs influence the spider to build again and with significantly more material and thicker

Table 9. *Increase of material and web components for intact webs as contrasted with cut webs. In both instances the spider could injest all of the material. The intact web, however, first had to be disassembled by the animal*

Days during which web was not destroyed	% Change in					
	No. of spiral turns	No. of radii	web radius	web nitrogen	thread length	thread thickness
1	+17.4[a]	+6.0	+9.3	+40[a]	+25[a]	+5
2	+20.0[a]	+6.0	+9.3[a]	+90[a]	+25	+30
3	+22.4[a]	+3.9	+11.2[a]	+75[a]	+28	+70[a]
4	+34.1[a]	+3.0	+15.0[a]	+90[a]	+45[a]	+70[a]
8	+38.2[a]	+9.1	+1.8	+217[a]	—	+140[a]

[a] These figures are significantly different from controls below the 1% probability level. Each group of spiders has its own control and cannot be compared with the next group.

thread than did the animals with the wad of material. The webs have larger diameter and more spiral turns when beginning from old webs. Such observations raise the question whether an animal is somehow able to evaluate the condition of the web, and on perceiving that it was left intact during the past 24 hours, makes better use of the favorable situation by constructing a larger and heavier trap.

At present nothing seems to be published establishing measurements of the consequences of web position and place selection for food catching. It is not even known whether a small number of signals determine the start from a specific place or whether this has to be regarded as a more complex procedure. The results of our experiments make measurements of web positioning particularly interesting for the future.

F. Effect of Drugs

In 1965 (WITT and REED, 1965) we summarized the effects of 23 different drugs on frequency of web-building, size, regularity, and shape in a single table. For the purpose of uniformity, much information had to be left out and much had to be simplified. Rather than repeating the table with addition of the few drugs which have been investigated since, we will describe the effects of each drug or group of drugs separately.

In the course of the drug experiments it was possible to confirm the original observation that the web test can be used to distinguish among their effects. However, experiments with drugs were carried out with different goals, with different species of spiders, and with different methods of evaluation, so that the separate discussions appear necessary.

Encouraged by the evidence of the sensitivity of web-building to pharmacological agents, the web test was applied to the search for foreign substances in the body fluids of mental patients (RIEDER, 1958; BERCEL, 1960; WITT and WEBER, 1956; WITT, 1958). One problem in these studies was the choice of human controls: urine, plasma, cerebrospinal fluid of control subjects may contain many substances which affect the web test. Follow-up experiments seem necessary in the cases where body fluids were found effective (RIEDER, 1958a; BERCEL, 1960) in distorting the web pattern.

D-Amphetamine

Confirming and amplifying earlier experiments by WITT (1949), WOLFF and HEMPEL (1951), WITT (1952), WITT et al. (1961) studied the effects of d-amphetamine on the web-building of spiders, alone and after pre-treatment with iproniazid and imipramine. The extent of behavioral changes was compared to the concentration of the drug in the body tissues, under the assumption that a higher concentration of d-amphetamine in the body (for example after pre-treatment with iproniazid) would generally correspond to more severe web distortion. WOLFF and HEMPEL (1951) had shown previously that methamphetamine affects 3 web proportions: the size of the catching area, the regularity of angles, and the regularity of the spiral. The last is the most obvious to the eye and the most difficult to measure and calculate. Since the experiments were carried out before the introduction of a computer evaluation, degree of abnormality of webs was rated by two observers, in effect using the integrative capacities of vision.

As expected, the web size was decreased the day after 600 mg/kg of d-amphetamine and 60 mg/kg of d-amphetamine 1 day after iproniazid-pretreatment, as well as 2 days after 60 mg/kg of d-amphetamine in spiders treated 3 days previously with 600 mg/kg of iproniazid ($P < 0.001$). Each drug of the above combination, when administered alone in the same dose, was ineffective. The change in size after 4 days was only significant at the 5% level, and after 8 days the figures were back to normal.

Both observers judged a majority of the webs to be grossly abnormal after 600 mg/kg of d-amphetamine as well as after 60 mg/kg d-amphetamine 1 and 3 days after iproniazid. Examples are shown in Figs. 20 and 21. The Chi-square test indicated a significantly increased number of abnormal webs, below the 1% probability level, after 600 mg/kg of d-amphetamine as well as after 60 mg/kg in iproniazid pretreated spiders. 4 days after iproniazid, 50% of the d-amphetamine webs were still regarded as grossly abnormal. 8 days later 20% were judged abnormal; 10—14 days later all apparently had returned to normal. In all groups there was good agreement between the two observers. Neither the (60 mg/kg) d-amphetamine webs after imipramine nor webs after iproniazid alone appeared grossly abnormal.

By contrast there was only one group of webs significantly changed after imipramine-d-amphetamine, and the direction of change was opposite to that observed after iproniazid-d-amphetamine. Webs were increased in size when 60 mg/kg of d-amphetamine were given 5 days after imipramine. This result is similar to that described by WOLFF and HEMPEL (1951) after low doses of methamphetamine.

Iproniazid in a dose which alone did not affect the regularity of angles, caused webs with significantly more irregular angles ($P < 0.01$) — similar to those after the high dose of d-amphetamine alone (P < 0.01) — when followed 3 days later by an ineffective dose of d-amphetamine. If d-amphetamine was given 1 day after

Fig. 20. A characteristic amphetamine web, built by a *Zygiella* 12 hours after 300 mg/kg of the drug, is smaller and contains less-regularly-spaced radii and spiral thread than do control webs. The effect on the web seems similar in several species (compare *Nephila* in Fig. 10b and *Araneus* in WITT et al., 1961)

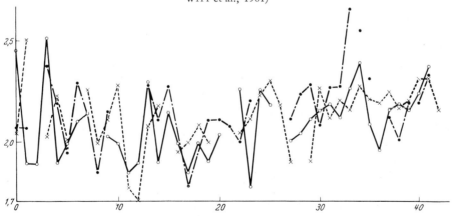

Fig. 21. Amphetamine effects on spiral regularity: Distances between spiral turns (ordinate) plotted against number of turns on one radius (abscissa). The variation in interspiral distances is unusually large

iproniazid, only a few webs were found to have highly irregular angles. This accounts for the large standard error of the mean after 1 day being significantly larger than after 3 days (P = 0.03). None of the changes following imipramine application was significant. The average figures for imipramine deviate in the opposite direction from iproniazid.

In measurement of body-drug concentration, spider-homogenates showed a decrease in monoamine oxidase activity after iproniazid and recovery of the enzyme in about 15 days. At the time of maximal MAO inhibition, the catabolism of d-amphetamine in spiders' bodies was slowed down significantly and the effect of d-amphetamine on spiders' webs had increased about tenfold. On the 5th day after iproniazid, the d-amphetamine sensitivity of spiders was again close to normal; d-amphetamine catabolism and MAO activity were only slightly slowed down. It is concluded that iproniazid inactivates the system catabolising d-amphetamine as well as the MAO system.

Web-building was apparently unaffected by both antidepressant drugs alone but changed whenever a sufficient concentration of d-amphetamine had been reached in the spider's bodies.

Mescaline and Psilocybin

Results of earlier experiments with mescaline (WITT, 1956) and psilocybin (WITT, 1960) were confirmed and enlarged, as well as the effect of mescaline compared with the effect of psilocybin by CHRISTIANSEN et al. (1962). Only female spiders *Araneus diadematus* or *Araneus sericatus* were used with no preference for age or weight. The experiments were performed all year round in the laboratory under controlled conditions. In order to establish the independence of the effects from seasonal variation in web-building, experiments with one drug and dose were repeated after 14 months. In each such double series the results were similar.

Drugs: Crystalline psilocybin from Sandoz Pharmaceuticals was dissolved in tap water and enough glucose added to make it taste sweet. Each spider was weighed and the number of drops and concentration of drug in the drops were adjusted so that each spider got the desired dose $\pm 10\%$. Mescaline hydrochloride crystalline was obtained from Hoffmann-La Roche Inc. It was dissolved and given to the animals in the same way as described for psilocybin.

The total thread length in the spider's web decreased significantly after psilocybin 150 mg/kg or mescaline 1 g/kg. For a dose of 250 mg/kg mescaline no significant shortening of the thread was apparent; a higher dose (500 mg/kg) showed measurable shortening, but the number of experiments was few. It appears therefore, that the shortening effect is dose dependent. The thread length decrease produced by 150 mg/kg psilocybin is comparable to that produced by 500 to 1000 mg/kg mescaline.

By the following day, thread length had returned to normal levels for both psilocybin and mescaline animals.

The short thread built by psilocybin animals was not thicker. The mean weight per m thread was even lower than controls, although the difference was significant only at the 2.6% level.

If we had measured only thread length, it might have been inferred that mescaline 1 g/kg and psilocybin 150 mg/kg caused equal web changes. However, while the webs built following psilocybin did not differ from controls in regularity of angles, irregularity became significant for the mescaline treated spiders. Tis result can be seen in Figs. 22a,b, 23, 24. A lower dose of mescaline (250, 100, or 50 mg/kg)

a   b

Fig. 22. a Control web of an adult female Zygiella and b web of same individual built under the
influence of mescaline. Note the reduction in size of the spiral-covered area

Fig. 23   Fig. 24

Fig. 23. Another mescaline web shows changes similar to those in Fig. 22
Fig. 24. A third mescaline web shows characteristic spiral changes and the wide free zone

caused no significant change in the regularity of angles, indicating that this effect depends on a dose in the same range as that effective for shortening thread. The webs after 1 g/kg mescaline were also significantly smaller than controls below the 1% probability level, while lower doses of mescaline showed no such effect. Psilocybin failed to affect angle regularity at any dose level.

As for web-building frequency, any dose of psilocybin above 150 mg/kg significantly decreased (P < 0.01) or totally suppressed web-building, but even 1 g/kg mescaline failed to have any effect. The few webs built after high doses of psilocybin showed no indication of irregularities in angles. If the effect on web-building frequency is considered, a psilocybin dose between 150 and 300 mg/kg was as effective as a mescaline dose between 1 and 5 g/kg. Both mescaline and psilocybin made the spider spin a shorter thread. Each drug has this effect in a different dose range. The relationship between the potency of mescaline to psilocybin is about 1 to 10 in spiders, psilocybin being the more potent drug. The result with respect to web-building frequency confirms this ratio (a mescaline dose exceeding 1 g/kg as well as psilocybin dose exceeding 100 mg/kg reduced frequency).

Measurements of thread length and web-building frequency alone would suggest that psilocybin and mescaline have similar effects in spiders. The method of testing the drug effects on web-building has, however, the advantage of permitting us to compare additional parameters. The web size as well as the regularity of angles is differently affected by both drugs, mescaline alone causing smaller and less regular webs. This indicates that mescaline interferes with motor behavior at a dose at which web-building is still possible, while psilocybin does not change regularity as long as webs are built at all. An interpretation that psilocybin interferes first with web-building drive before it affects motor behavior was suggested by CHRISTIANSEN et al. (1962). Mescaline, in contrast, would interfere with motor coordination first and resembles in that respect the effects of d-amphetamine on web-building.

Web-building recovers from the effects of both drugs on the following day.

Caffeine

In 23 experiments *Zygiella-x-notata* Cl. spiders received dosages up to 100 µg caffeine per animal (corresponding to 1 g/kg for a 100 mg spider) in sugar water by mouth 8 hours before web building time; 256 webs built without drugs were used for comparison (WITT, 1949). There was no significant change in web-building frequency or regularity of spirals. There was, however:

1. A statistically significant increase of the relative length of the horizontal parameter of the catching area as compared to the vertical (Webs were more round; compare with strychnine effects);

2. A significant (P < 1%) increase in the number of oversized central angles;

3. A significant (P < 1%) increase in the number of radii which did not run full length from the hub to the frame, being somewhere disrupted.

4. A 5% significant decrease in the size of the catching area.

After application of the highest dose, the changes persisted for 2 days (32 hours and more). Very severely disturbed webs are shown in Figs. 25 and 26. In these experiments building time had more than doubled. The severely disturbed webs can hardly be measured and constitute altogether borderline cases of web-building. It

Fig. 25. The *Zygiella* web built after a high dose of caffeine contains hub, radii, frame, spiral and signal thread, recognizable but built in a most disorderly fashion

Fig. 26. Another web shows characteristic caffeine effects as well as idiosyncratic changes when compared to that in Fig. 25

is not only surprising that these obviously severely handicapped animals built at all, but that a hub, radii, spiral, frame and the characteristic free sector of the *Zygiella* web are still recognizable.

It can be safely assumed that higher dosages of caffeine disturb the motility of the spider severely while not interfering with the drive to build. Of interest are the interrupted radii. Was a change in silk production or handling responsible for these interruptions? (see also xylopropamine, WITT, 1956, and chapter 4).

Strychnine

WOLFF and HEMPEL (1951) in 91 experiments with *Zygiella-x-notata*, gave strychnine salt, strychnine-N-oxide or "Movellan-acid-ASTA" by mouth in sugar water. They tested doses from 10—82 μg per animal (corresponding in a 100 mg animal to 100—820 mg/kg) and found doses of 30 μg and more per animal effective. The changes were as follows:

1. Web building frequency was distinctly diminished.

2. The spiral area was smaller.

3. An analysis of the shape of webs in terms of fit to one of four types, established increased frequency of round webs with a tendency toward decrease of pendulum turns in the spiral and relative increase of full turns.

The original paper illustrates the "strychnine-shape" in a number of photographs and should be consulted. The controversial, but frequently described effect of strychnine on sensitivity (see for example POULSSON, 1923) is generally ascribed to a lowering of the threshold of nervous reflexes. This interpretation could be applied as explanation for the changes in web pattern (WOLFF and HEMPEL, 1951). Clarification of such a concept as well as additional experimentation with spiders must, in our opinion, precede further speculation.

Phenobarbital (Luminal) and Diazepam (Valium)

The experiments were undertaken to study two characteristics of drug action, time- and dose-dependence, in web-building (REED and WITT, 1968). It was assumed that by increasing dose, and by timing building for the peak effect of the drug, the physiological mechanisms of web construction could be progressively disturbed. The experiments were also carried out to test reproducibility of method with a large number of spiders in different years, repeating the same procedures 12 months apart. Such an investigation demands a great number of data and calculations. These experiments were done with the aid of the computer; 27 parameters of over 1000 webs in groups of 20 spiders were compared. The two drugs seemed of interest because they both have sedative properties but are chemically and therapeutically distinct. Phenobarbital is known as one of the long acting barbiturates with special anticonvulsive effects, while diazepam is classified as a tranquilizer with muscle relaxing effect and delayed onset of action. Diazepam also is the more potent drug, being given to man in 1/100 of the dose of phenobarbital.

Web-building was tested between 12 and 48 hours (and occasionally 60 hours) after drug application at 12 hour intervals. The doses were 10, 100, 200, and 400 mg/kg of diazepam and 10, 100, and 1000 mg/kg of phenobarbital in sugar water by mouth;

the amount was adjusted to the body weight of the spider. The body weight varied between 6 and 800 mg.

The following changes were observed after diazepam: curtailment of expenditure of thread occurred for all dose-levels used in the experiments. There was a significant reduction in the total length of the lines composing the web, lasting at least through 36 hours. The curtailment of thread seemed to be a matter of reduction of number of thread elements rather than shrinkage of the space occupied by the web; the threads of the main web, forming the viscid or catching zone, were consistently fewer in number in the drug webs, while the area covered by the spiral showed only occasional reduction in magnitude. Radial threads declined significantly in number after diazepam doses above 10 mg/kg: the effect never appeared in the 12 hour webs and was absent in the 48 hour webs for the 100 mg dose level and at 60 hours for the 400 mg dose level.

The nitrogen per web decreased from a control level of 35 units to 21, 20 and 22 units respectively for the different post-drug periods. This means that the webs built under diazepam contained less material than did control webs. The continuous decrease in thread length and in material indicate that there was no thickening of the thread as a net effect.

In order to determine whether diazepam had a direct effect on silk synthesis, the amount of thread which could be pulled from the ampullate glands on control days was compared with the amount of silk which could be pulled one and two days after application of 100 mg/kg diazepam. There was no change in the amount of silk which could be pulled, indicating that it was the expenditure rather than the production of the thread which was affected by the drug.

Statistical tests for departure from regularity produced significant and near significant angle irregularities for several experiments with diazepam, but clear irregularity only at 200 mg/kg or more. Spiral regularity stayed at the normal value.

The phenobarbital webs were significantly reduced in thread length, number of spirals, number of radial threads and spiral area at all dose levels. These effects lasted through 36 hours. Irregularity of central angles appeared in the lowest experimental doses and the earliest constructions; it persisted throughout 36 hours. Spiral irregularity appeared only in the webs of the high-dose level, 1000 mg/kg.

It is concluded that both diazepam and phenobarbital produce a significant diminution of the activity level of the spider without affecting the amount of material available for the web. The decline in activity is more radical for phenobarbital than for diazepam: disturbances which appear early and at low doses of phenobarbital are present only in the late and high dose-levels of diazepam. However, there appears to be an order in which the radial and spiral threads are affected by the drugs: a faster "penetration" into the mechanisms underlying placement of radial lines than those underlying spiral-building. Spiral irregularity, which is present for the high doses of phenobarbital does not appear in the diazepam records; whether it may do so at higher doses of diazepam cannot be answered from these experiments.

Phenobarbital produces a decline in level of activity and in sensory-motor control, the latter being only intimated at high dose levels of diazepam. An order of vulnerability to disturbance of web-building is suggested by the time and dose-dependent results:

1. There is curtailment of activity, where the available silk is not emptied from the gland, as appears to be the case in normal building. The total path, as measured by length of thread, is reduced without loss of precision of placement of threads and without reduction in the area of the spiral zone.

2. The dimensions of the spiral zone are contracted. Precision of placement of threads is still preserved, but because of the smaller area of the essential sticky trapping region, the web is less effective as a snare than the normal web.

3. There follows a loss of precision of placement in two substages (a) the central angles of the web become more variable, less regular than normal, (b) spaces between spiral turns become less regular than normal. (After phenobarbital the angle irregularity distinctly precedes the spiral irregularity. At the dose levels employed, diazepam did not cause detectable change in the spiral regularity measure.)

It may be concluded that degree of web-building disturbance is positively correlated with dose level and with time of drug action; but an inverse relationship in potency of the two drugs is found for web-building disturbance and therapeutic effects in man.

Chlorpromazine (Thorazine, Largactil)

WITT and HEIMANN (1954) 19 and 10 hours before web-building time gave single doses of 1, 10, 25, and 100 µg/animal chlorpromazine (corresponding to 10 mg to 1 g/kg) by mouth in sugar water to 51 *Zygiella-x-notata* Cl. There was no observable

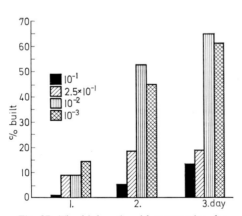

difference in effects between the early and late time, and the web change was related to dosage alone: Fig. 27 shows that with increasing dose, web-building was longer and longer interrupted, the difference between frequency at the highest and lowest dose showing a Chi square figure of 5.56 with $P:0.01 = 6.6$. As long as any webs were built, their parameters were undisturbed. This finding is confirmed by GROH and LEMIEUX (1964).

Only one animal was observed by WITT and HEIMANN to build a highly irregular web; it occurred on the second day after a high dose of chlorpromazine. This web resembled those shown by GROH and LEMIEUX (1964) which were built after 50 mg chlorpromazine per day when the drug was administered during 20 consecutive days. Under these circumstances smaller webs with a decreased number of radii were observed; the radii were also found "less elastic". The changes lasted throughout the spiders' life-time. Type of web disturbance and the slow movements of chronic chlorpromazine spiders lead the authors to compare their results with the behavior of old animals. However, the duration of the spiders' lives was not diminished.

Fig. 27. The higher the chlorpromazine dose (concentration in each drop indicated by differently designed bars), the longer (abscissa, days) the period of reduced web-building activity (ordinate, %)

If one considers that some kind of web is still constructed after most severe disturbances, the complete interruption of web-building after single applications of chlorpromazine can be interpreted as the drug interfering with "drive". Results of the chronic experiments could be explained in a similar way, assuming also that irreversible damage to the central nervous system substrate had occurred as a result of the drug.

Another tranquilizer, trimipramine, produced even more severely disturbed webs in experiments by GROH and LEMIEUX (1964). After a single dose of 30 µg per animal webs were smaller, had fewer radii and were more triangular in shape. 5 days later these spiders were again building regular webs. However, repeated administration of this drug lead to irreversible changes. The authors interpret these results, particularly the shape change, as a regression to the phylogenetically earlier web form of, for example, Hyptiotes (Fig. 13).

D-Lysergic Acid Diethylamide (LSD 25)

Pure substance (Sandoz) was given by WITT (1951) by mouth in sugar water in two dose ranges to *Zygiella-x-notata*: the higher dose of 0.1—0.3 µg per animal (corresponding to 1—3 mg/kg in a 100 mg spider) caused the following changes in 13 experiments:

1. longer webs in 88.7% of all webs;

2. possible smaller catching area and possibly more irregular angles (5% significance).

And no other measurable changes were found.

The lower dose of 0.03—0.05 µg per animal (corresponding 0.3 to 0.5 mg/kg in a 100 mg spider) produced characteristic changes which are rather surprising:

1. There was a decrease in web-building frequency of 74.1% ($p < 1\%$ significant).

2. An increase in angle regularity ($< 1\%$ significant).

3. An increase in spiral regularity ($< 1\%$ significant).

4. An increase in catching area size and a decrease in the number of oversized angles (but significant only at the 5% level) (Fig. 28).

A first interpretation assumed an increased speed of association through improved evaluation of sensory input. In a later paper (WITT, 1952) a tentative explanation of spiral building was proposed which would explain the LSD effects as a decreased interference by "disturbing" steering components during spiral construction and consequently the more undisturbed orientation to the shortest path from radius to radius. Motor control would be unaffected.

Physostigmine

After the webs had been observed for two preceeding days, 1 mg/kg physostigmine was given by mouth to 24 spiders 12 hours before their normal web-building time. Table 10 shows that there was no significant difference in total web nitrogen and web measures between the control day and the first day after physostigmine. This result is in agreement with the gland data (PEAKALL, 1964a) which indicates that a physostigmine effect on silk production could hardly be visible in the webs after such a short time.

Table 10 shows in the third column that the webs which were built 36 hours after physostigmine contained 32.5% more nitrogen than on the preceeding day. The webs were measured without being sprayed in order to allow the spider to eat the

Fig. 28. Reduced variance in central angle size and spiral spacing occurred in webs built after a low dose of LSD 25

Table 10. *Means and standard deviations of measures taken from 20—24 webs*

	Control webs	1. Drug day	2. Drug day	3. Drug day
Nitrogen/web	29.1 ±16.7	36.9 ±18.8	48.9 ±21.5	34.5 ±18.3
Thread length (m)	14.25± 4.68	12.81± 4.11	16.44± 5.05	13.83± 5.12
Thread thickness	2.83± 1.18	2.87± 1.09	2.97± 0.99	2.59± 1.12
No. of radii	27.5 ± 4.1	26.3 ± 4.2	27.8 ± 3.3	27.0 ± 3.7
No. of spiral turns	32.5 ± 8.2	29.3 ± 6.7	35.3 ± 7.7	31.3 ± 8.2
Mean radius of catching area (mm)	89.6 ±16.1	87.5 ±15.5	95.5 ±18.3	88.6 ±20.7

1 mg/kg physostigmine was given by mouth on the evening of the day on which control webs were built. The first drug webs were built 12 hours after drug application, the second 24 hours after the first.

Observe the slow onset of drug effects (2nd drug-day) and the way spiders redistributed the increased amount of silk on thread length rather than on thickness. On the 3rd drug-day all measures were back to control values.

thread; changes in web proportion could be established only roughly. However, we have a clear indication in the data of how the animals distributed the larger amount of thread-polypeptide: the thread did not become thicker (that is, there was no change in body weight) and the number of radii did not increase significantly under the influence of physostigmine; however, a larger web ($+9.6\%$ mean radius) with a longer thread ($+28\%$) and more spiral turns ($+31\%$) was built.

The figures for the following day indicate that there was a return to control values by 60 hours after drug application. No rebound effect in silk quantity occurred. A group of 12 control animals, kept and fed in the same way during these days, but not receiving physostigmine, proved through lack of change on the days corresponding to the second drug day that the alteration occurred only in the drug group.

These data together with the measurements obtained on isolated glands and on the speed of incorporation show the effect of a cholinesterase-inhibiting drug on thread production. In addition, we learned that the spider adjusts the web pattern to the increased amount of silk. While the thread thickness appears not to be increased, the animal has a longer thread length at its disposal and makes a larger, closer knit web. Such experiments show that there is a close interaction between gland filling and web pattern regulation. In addition they point toward a mechanism of silk regulation which may very well be under central control.

Atropine

1, 2 or 4 mg/kg atropine sulfate were given to 19, 19 or 39 spiders by mouth 12 hours before web-building time (WITT, 1962). The two lower doses caused webs which showed no change in size or regularity but were built with wider meshes, covering the same area with less thread. The highest dose caused significantly smaller and less regular webs built with less thread. The change lasted through the second day after drug application. It was suspected that in addition to a central nervous system effect, atropine might interfere with thread protein production in spiders, comparable to the reduced thread length and thickness which was observed after atropine in silk worms (TAMANO and KURIAKI, 1961). A method was devised to pull the thread produced by the spiders during the last 24 hours and weigh it. 4 mg/kg atropine reduced the amount of thread which could be pulled by 58%. This finding, together with other evidence from dissected spiders points toward the silk gland as the point of attack for atropine in interfering with thread production. Compare these results with those discussed in the chapter on regulation of protein synthesis in silk glands.

Scopolamine

The effects of scopolamine on web-building of *Zygiella-x-notata* Cl. was investigated by WOLFF and HEMPEL (1951) in 60 experiments. They administered 5—60 µg per animal (this would correspond in a 100 mg animal to 50—600 mg/kg). In some instances they repeated application at 24 hour intervals for cumulative drug effects. The degree of disturbance increased in the webs with increased dose or repetitive application of the drug. Effects comparable to those found by WOLFF and HEMPEL after scopolamine were also found in spiders' webs built after feeding urine of a patient who had received 0.75 mg scopolamine the day before (WITT and WEBER, 1956).

5*

No change in web-building frequency was observed. However, WOLFF and HEMPEL found changes in the following web-parameters:

1. There was a decrease in the size of the catching area.

2. Webs were relatively elongated in the direction of gravity.

3. Central radial angles were more irregular, radii were loose and distorted, and the number of oversized angles was increased.

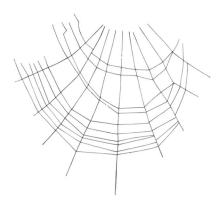

Fig. 29. After scopolamine the spiral is spaced in a characteristic fashion, deviating periodically far from its normal path, as indicated in this drawing made from a web photograph. Note the difference between amphetamine and scopolamine spirals

4. There was a systematic disturbance in the course of the spiral which led sometimes to more than one center (Fig. 29).

5. Comparing 72 drug webs with 158 control webs and using the special measurement in Zygiella webs of distances from retreat to hub over hub to opposite end of catching area, an increased variation in the position of the hub was found. The catching area was sometimes completely below the hub.

In contrast to the effects of pervitin, these changes are interpretable as consequences of a disorientation which becomes apparent first in the central parts of the web. Whether this is based on sensory misinformation, motor failure, or disturbances in central information processing, cannot be decided. The periodicity of spiral deviation may point toward a disturbance appearing at short intervals, or a regulation of spiral regularity which sets in only belatedly after the deviation from the regular spiral course has become excessive.

Diethyl Ether $(C_2H_5)_2O$

Low concentrations of ether applied as pads of cotton wool soaked in the drug were placed on top of a plexiglass cylinder which contained the spider frames. Three experiments before building time prolonged the web-building time but showed no other changes (SCHWARZ, 1956).

Higher concentrations which were sufficient to anesthetize spiders in 1 minute proved harmless, if kept up for 12 hours. When the atmosphere was saturated with ether above the point of the just anesthetic concentration, eight animals were killed in the space of 1 hour.

SCHWARZ performed a number of single experiments on different types of web-building: one spider was put into a saturated ether atmosphere at the start of sticky spiral building: there were visible changes in web geometry, the spider missed threads, several times nearly fell out of the web, and the time to finish the spiral was 90 minutes.

In another instance a spider was put into the ether atmosphere at the end of the building of the auxiliary spiral: the animal tried to return to its retreat, and missed the direction. It built finally a few irregular spiral threads.

Another *Zygiella-x-notata* received ether inhalation at the beginning of the sticky spiral: it interrupted for a while, recovered 30 minutes later, and tried for several minutes to find the retreat.

One spider built in ether atmosphere in the course of 20 minutes a few wide spaced spiral turns, but stopped before it reached the end.

Another spider stopped building immediately upon being brought into the ether atmosphere.

A last spider was anesthetized and stopped moving before it finished the radii. 14 minutes later, when removed, the spider woke up and built one more radius, showed signs of disorientation and built two spiral turns in the course of the following 90 minutes. The spiral turns cross over each other, diverge and converge. The auxiliary spiral is only partly eliminated. Radii are loose. Even the free sector of the Zygiella web is crossed with sticky spiral turns.

The spiders showed behavior not observed under nitrous oxide; twitching of the legs, rubbing of palpi and first legs, and much longer times to complete the spiral after ether.

It can be taken as a generalization that the effect of ether by inhalation in spiders is of rapid onset and, after a certain period, is completely reversible as far as web geometry is concerned. The effective concentration, though difficult to measure, seems to be similar to the anesthetic concentration in man and higher animals. We have not observed any suffocation of spiders in ether atmosphere. The spiders showed an introductory stage into anesthesia with excitement and twitching, disorientation, loss of equilibrium and direction, like higher animals. Spiders never fell completely out of the web, but sometimes seemed close to this. The webs looked different from those after amphetamine, strychnine, and looked similar for different spiders. SCHWARZ comes to the conclusion that the spiders after inhaling ether for a short time show something similar to the excitation stage of anesthesia in man followed by deep, reversible anesthesia.

Carbon Monoxide (CO)

To study problems of chronic poisoning, development of tolerance, and to compare lethal dose in spiders to that in insects, young and adult *Zygiella-x-notata* were exposed to 100% coal gas (as used for lighting) by EPELBAUM (1956). The composition

Table 11. *Effects of single and repeated inhalation of carbon monoxide*

No. of animals	In 100% coal gas for mins	Repeated on consecutive days	Result
2	60	once	1 dead, 1 recovered
6	45	once	2 dead, 4 recovered (normal web-building)
3	30	once	all alive and normal
8	10	once	restless, weak movements, recovery in 3 mins, no web changes
4	10	10	increasingly quick recovery, normal growth, webs unchanged
3	10	6	no change in webs
4	10	5	quick recovery, after incoordination and motionlessness

of gas was 16.7% CO; 4.6% CO_2; 0.7% O_2; 50% H_2; 17.9% CH_2; 2.0% C_nH_n. The exposure lasted from 5 to 60 minutes, and was carried out either 1 or several times, up to ten times, once a day. Table 11 shows the results.

The sequence of effects was: restlessness, disorientation, followed by completely reversible anesthesia. Repeated applications at 24 hours interval lead to development of tolerance rather than to hypersensitivity. No lasting after-effects were observed.

Carbon Dioxide (CO_2)

Six spiders were kept in a 40—100% CO_2 and O_2 atmosphere for 10—20 minutes one or two times, 24 hours apart. There occurred anesthesia followed upon recovery by undisturbed web building (web measurements of size, shape and regularity were taken by EPELBAUM, 1956). These results contrast with those obtained by VAN DER KLOOT and WILLIAMS (1954) on silk worms, where striking changes in behavior during cocoon spinning were observed after CO_2 without changes in body development, MACKENSEN on bees (1947), where queens precipitated oviposition, and SIMPSON (1954) where workerbees showed deep and lasting behavior changes. Such differences between the two insects and the spider could originate in bodily chemical differences or in a dissimilarity in the way in which the gas was applied to the animals. It should be easy to test this experimentally with other anesthetic drugs on the three species.

Nitrous Oxide (N_2O)

SCHWARZ (1956) enclosed the frames of *Zygiella-x-Notata* in tight plexiglass cylinders before or during web construction. The enclosure permitted different concentrations of gases to be produced in a short time with a minimum of air currents. A spider supposedly inhales such gases through the spiracles which open on each side of the epigastric furrow into the book lungs and through its second, later evolved, respiratory system of tracheae which is entered through a single spiracle just in front of the spinnerets.

The anesthetic dose, which depends on concentration, atmospheric pressure, and duration of exposure, was found similar to that established by animals at 90—100% N_2O after several minutes at normal atmospheric pressure. Death occurred relatively late, after 20—24 hours in 100% N_2O, showing a possibly high tolerance of spiders to lack of oxygen. Recovery from complete anesthesia took about 30 minutes in oxygen and was complete. Only normal webs were built on the following day.

While no webs at all were constructed if the animals were kept for 12 hours before web-building time in a 80% N_2O + 20% O_2 atmosphere, 8 spiders showed interesting behavioral changes when N_2O was applied just before the start of the sticky spiral: four spiders interupted building, became disoriented, did not find their way back to the hiding place, and finally hung motionless in the web, usually in the hub. Four other spiders showed signs of severe disorientation manifested in spirals crossing over or running apart, in uncoordinated movements resulting in loose threads and closing of free-sector, lack of elimination of the auxiliary spiral and a premature end to building.

SCHWARZ concludes that an excitation stage precedes anesthesia in spiders as in higher animals and that the gas has a central nervous system point of attack rather than suffocates spiders. His sample pictures of webs finished under N_2O influence demonstrate the erratic route of the web builder, resembling the effects of high doses of caffeine rather than amphetamine webs.

Adrenochrom

10—40 µg Adrenochrom per animal were given to 16 *Zygiella-x-notata* Cl. by Witt (1954). 13 spiders died immediately. 4 mg adrenochrom per animal given 6.5 hours before web building time to 23 *Zygiella-x-notata* spiders did not change the frequency of web-building, but caused the following web pattern changes:

1. Highly significant decrease in the size of the catching area.

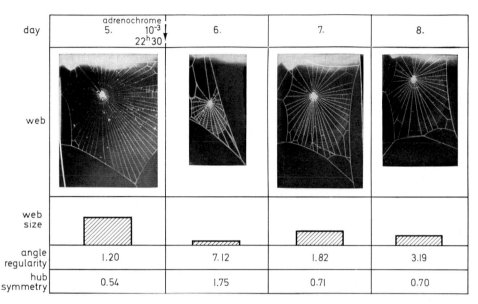

day	adrenochrome 5. 10⁻³ 22ʰ30	6.	7.	8.
web				
web size				
angle regularity	1.20	7.12	1.82	3.19
hub symmetry	0.54	1.75	0.71	0.70

Fig. 30. The figure illustrates an experiment with one spider, and the effects characteristic for adreno-chrome. On the left, the photograph shows the control web and below three sample measures; the second photograph from the left is of a web built 24 hours later, about 8 hours after drug application; the two figures on the right follow the first by 48 and 72 hours and are examples of webs built after recovery

2. A 5% significant decrease in the regularity of the angles.

3. Highly significant increase in the standard deviation of hub symmetry.

The change in hub symmetry was reverted to control values on the following day while the other two changes stayed statistically significant for another 24 hours. In contrast to scopolamine which caused similar changes, adrenochrom did not influence the regularity of the sticky spiral.

Figs. 30 and 31 show the characteristic sequence and consistency of changes of an adrenochrom experiment. We have at present no interpretation of the changes in web patterns as the result of disturbance in the central nervous system.

A solution of adrenochrom changes quickly through oxidative processes in the air. Different laboratories have reported different results with adrenochrom solution which can probably be attributed to this deterioration (Bacq, 1949). In spiders the

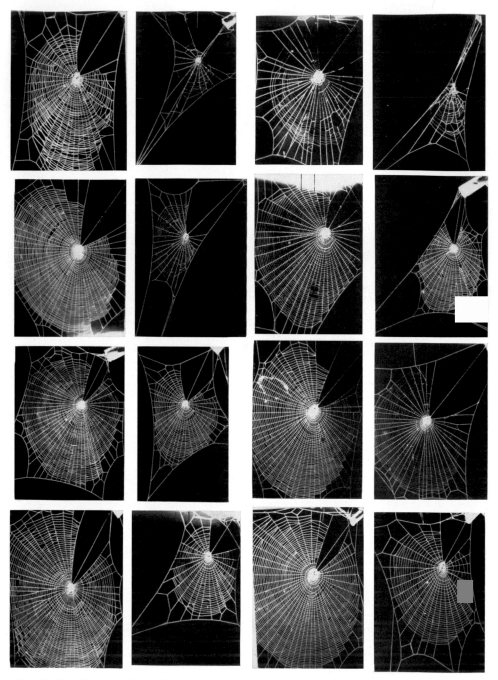

Fig. 31. The first and third column show photographs of *Zygiella* webs constructed the morning before drug application, the second and fourth column show the first web of the same spiders built after adrenochrome application. Note similarity of drug webs

experiments with the fresh adrenochrom solution were compared with another group of experiments where adrenochrom solution was used which had stood for 24 hours in an open glass and where the bordeaux red color had changed into an opaque brownish-red.

35 spiders showed no lethal effect after ingesting 10—40 μg of "old" adrenochrom solution per animal. After 4 μg per animal there was no measurable influence on frequency of web-building, however, 6—8 hours after drug application the following changes were observed:

1. Highly significant decrease in the size of the catching area.

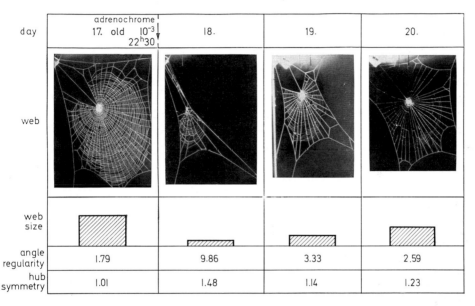

day	adrenochrome 17. old 10⁻³ 22ʰ30	18.	19.	20.
web				
web size				
angle regularity	1.79	9.86	3.33	2.59
hub symmetry	1.01	1.48	1.14	1.23

Fig. 32. Web changes produced by an old adrenochrome solution, which had altered its color, indicative of a condition which has been reported to induce hallucinogenic effects in man (HOFFER et al., 1954) are indistinguishable from those after a fresh drug solution

2. Highly significant decrease in angle regularity.

3. Highly significant increase in the standard deviation of hub symmetry.

All these changes (with the exception of the decrease in catching area) are not measurable if the solution has been given 11—13 hours before web-building time (WITT, 1954). The characteristic changes of the web under "old" adrenochrom are best shown in a number of photographs (Fig. 32).

We can summarize the changes in the following way: adrenochrom causes very small webs which lie rather far away from the retreat of *Zygiella-x-notata* and possess unusually long signal threads. The upper part of the webs (frame thread and upper spiral turn) lie relatively low in regard to the hiding place. The web has a small number of radii and the sticky spiral is nearly exclusively on the side which is opposite the retreat. The effect lasts only a few hours, certainly not more than 10 hours. The "old" solution shows less of the lethal effect, but causes similar web pattern changes as the freshly dissolved adrenochrom.

V. Construction of the Web

Although the course of events in the building of the orb-web have been carefully observed and described, for instance by SAVORY (1952), the sensing and executing mechanisms which serve the process remain subjects of speculation. It is clear that there are episodes which introduce and terminate phases of building (KOENIG, 1951). There are evidences of imperatives stemming from the internal condition of the spider as well as signs of guidance by external circumstance. Some stimuli result from previous events in construction which have led to a particular array of threads in space; some follow progressive depletion of material in the silk glands. These stimuli interact with those provided by certain system conditions: at one level, the weight of the spider and the length of its legs, at another level the central program guiding construction.

These relations between stimuli and behavior have been inferred from observation and experiment on the web; they have led to curiosity about the information flow which inaugurates and regulates the program and which is presumably based upon neural and sensory elements acting in intricate concert. As in most animal behavior, there is a large gap between description of the spider's complex behaviors and knowledge of the sensory and motor apparatus which underlies its behaviors. Despite some informative work on sense organs and some presumptive relations between anatomy of the central nervous system and complexity of the web, the linking of the two bodies of information is a matter of program rather than of achievement.

However, the prospect of achievement receives some encouragement in the case of the orb-weaving spider — not because of the zeal of investigators, but in the possibility of objective analysis of suitably complex behavior. Usually, when precise specification of behavior is possible, the events themselves are elementary responses of sensory and motor systems, not the organized behavior of the whole animal. On the other hand, such organized behavior of the whole takes place against contingencies vast and complex. In the first case only sub-programs are treated. In the second case, the behavior is so complex that the language of programming seems to lose appropriateness. What, for instance, is the human nervous system programmed to achieve? That question requires elaborate reply not only for human beings but for most animals until rather simple levels of behavior are reached. It may be answered rather directly for the spider. The spider nervous system is programmed to achieve construction of a web through the spreading of thread. The execution of this program is not simple, but it does take place under contingencies which are relatively manageable for the investigator. The spider suspended from a few threads at the start of building has entered into a state of relative isolation in which the threads and their disposition are the animal's chief if not sole vehicles of information. They are literally its lines of contact. Presumably mechanoreceptors play a predominant role in the sensory input which then guides web construction, and the emerging structure of the web is a response to that input.

This chapter will review the geometry of the web of *Araneus diadematus* Cl. and sequences of building processes. Clues to the system of sensing and executing mechanisms will be sought in this record. Current knowledge of the anatomy and physiology of the nervous system of orb-weavers and their relatives will be examined for instruction concerning known structures and systems.

Table 12. *Characteristics of webs built in laboratory cages by 103 specimens of Araneus diadematus Cl. of various weights and leg lengths*

Web feature	Mean value	Standard deviation
Number of radii	26.08	4.24
Number of spirals:		
West[a]	20.45	7.50
North	21.42	8.24
East	25.58	8.40
South	29.63	10.04
Width/length	0.83	0.12
Area of spiral (cm^2)	328.28	189.80
Thread length (m)	14.33	6.72

[a] West is arbitrarily selected to refer to the side where the hub is closer to the frame.

The web of *Araneus diadematus* is an elastic, essentially two-dimensional cartwheel formed in a vertical plane (Fig. 1). It is composed of evenly-spaced radial threads or spokes, each running from an above-center hub to a peripheral framework of threads. Some of the radii bifurcate into Y-structures before reaching the frame; one arm of the Y-structure may be attached to an adjacent radius.

A distinction is sometimes made between framing and mooring threads, the frame referring to lines to which the radial threads are attached and the mooring referring to those lines which fasten to some non-thread structure such as a branch or the side of a cage. This distinction is useful descriptively, though the mooring thread may be continuous with the frame thread, having been placed by the spider in the same operation. The web proper may be thought of as a polygon formed by mooring threads pulling outward at the corners of the web; the vertex of the angle formed by the frame thread is pointed away from the center of the web. The effect of the radial threads, on the other hand, is centripetal. Their attachments produce a gentle or a sharp segmented curve in the frame thread between the mooring threads. Radial and mooring threads never meet at the same point; the spider does not make an attachment where one already exists.

In the complete web, the catching-zone is formed by a more-or-less continuous spiral of tacky thread which winds from the periphery toward the hub. It stops short of the hub, leaving a characteristic "free zone", that is, an area with only radial threads spanning it. The catching-zone is usually oval in shape, the long dimension in the vertical; this form is the consequence of the spider's frequent reversal of direction in laying the spiral, an event which occurs more frequently in the lower outer region of the web than elsewhere. The variability in the number of these two elements, the radii and spiral turns, is summarized in Table 12.

The central angles formed by the radii are regular in the sense that they are approximately equal in size, though those in the upper portion of the web tend to be larger than those in the lower. This uniformity is less apparent in the completed web than it is just at the close of radius construction. At that time the radial threads are taut and straight. They are deviated by the attachment of the provisional spiral. This thread, which winds from the hub and is attached from radius to radius, leads to the near outer edge of the web in a few wide spiral turns. It appears at the termination of radius construction, and provides a scaffolding and supporting element which does not appear in the completed web but is removed by the spider in the course of building the permanent, sticky spiral.

The permanent spiral is built from the periphery inwards, each turn being placed very close to the preceding turn. Under normal conditions the placement of these threads with respect to each other is highly regular. They lie closely parallel across the central angles between pairs of radii, and are best observed in the pristine state of the web when the spider has terminated the spiral and crossed the free zone into the hub. Later excursions to secure prey glue the viscid threads together.

The structural elements (hub, radii and frame) are composed of dry thread drawn from the ampullate glands as described in Chapter 2. The permanent spiral is composed of thread coated with material provided by the aggregate glands. While the viscid threads readily adhere to each other if they touch, the structural threads can be brought into a bundle and then separated; they do not coalesce although they will sometimes cling together. Wherever, structural threads intersect in the web they are joined by the spider. If the animal misses an attachment — a rare occurrence in *Araneus* — the threads can be observed to slide past each other if the web is moved. In the normal case, as a result of these attachments, motion in one part of the web is readily transmitted to all other parts. When alarmed or when trying to locate prey which is buzzing somewhere in the web, the spider may pull violently at the threads near the waiting position at the hub. The web contracts and expands in a few elastic pulsations in response to these tugs.

In order to produce the attachment at thread-intersections, it seems necessary for the strands to be carried to the spinnerets. While this maneuver may be observed in radial construction, it is most clear in the placing of the spirals.

A. The Process of Construction

Although it is a reasonable preliminary hypothesis that preset programs determine the course of web-building, even casual observation and reflection confronts the necessity to infer some mechanism for sampling concurrent conditions of the structure. The building operation encounters certain contingencies, first those concerned with the position of objects in space (which are to serve as supports) and second those presented by the existing structure at different stages of construction. Especially in the very early stages of web-building, when the animal seems to be groping toward some kind of rudimentary structure, lines are placed, shifted, removed, lengthened or shortened. In all of these maneuvers, the old line is detached and rolled up while the new line is pulled from the spinnerets. The preliminaries to construction seem to involve expendable or trial structures.

The building of the web is marked by phases. It is difficult to establish the onset of building or to define it except retrospectively. In or out of the web, immobility is the most usual condition of the sedentary web-builder, and the observer ambitious to follow construction from the very beginning must learn to tolerate long hours of uneventful watching and view skeptically certain business-like excursions of the spider which enliven the vigil. These brief periods of scrambling and thread-pulling are not necessarily productive of a web; they have an exploratory character, as if the animals were staking out space for the enterprise. The section on Methods above has described the conditions in which spiders are kept in the laboratory. The cages provide highly suitable structures for suspending webs, but they are explored anew for each web, providing the previous web has been destroyed by the experimenter. Between explorations, the spider lapses into its characteristic inverted crouch in an upper corner of the cage.

One apparently exploratory movement is a head-downward descent from a perch, dangling from a dragline. One of the rear legs grasps the thread; the others are extended. In a gentle current of air, the body of the spider turns about the thread, and a second thin thread emerges from the abdomen and is wafted in the breeze. The small thread is difficult to see, and it is not certain that it is always present, but it seems to be the precursor of a bridge-thread, emitted in search of a distant mooring. However, the fall of the spider could serve functions other than this apparent exploratory one. One possibility is that the thread material in the ampullate ducts requires renewal, and is therefore pulled out, ingested and replaced. At any rate, having stopped its descent just short of the floor, the spider, after a brief moment at the end of the dragline, climbs rapidly back to its perch, where it eats the line and lapses once more into immobility.

Immobility can also be provoked during web-building by alarming the spider. Vulnerability to alarm seems to lessen as the work of building progresses; at the same time, activity becomes progressively rhythmic. Although it is possible that sensory thresholds or other bodily conditions are systematically changing with the course of building, this invulnerability is probably simply a consequence of the multiplication of threads in the web. If a thread is gently parted by a hot wire during an early stage of building, when the web consists of perhaps five or six radial threads, the change in the tension state of the structure is more drastic than if the thread is burned when many supporting lines are present.

There is another phasic aspect to building which seems to depend on the existing structure and upon the internal condition of the spider. The period of radius-placing can be extended artificially by burning radial threads as they are built. KOENIG (1951) prevented orb-weavers from completing normal radius construction by removing all radii built when the spider was within a few lines of completion. He also destroyed all radii built by some spiders after the first three lines had been built. In both cases, the animal continued to replace the radial threads until it had worked through the equivalent of two to three times its normal complement of radial lines. If sensory feedback were the only stimulus to radius construction, this process of renewal would go on until thread material was lacking. However, the spider abandons replacement well short of this event (although occasionally the whole structure is abandoned) and moves on to the next stage of building. It begins to place the provisional spiral, even on a grossly inadequate scaffolding, e.g. the three-thread structure.

B. The Course of Construction

If no web has been left in the cage from the previous day, the nocturnal wanderings of the spider may have produced a few interconnected lines such as those shown in Fig. 33. An early morning hour is the usual building time for *Araneus diadematus*, but the presence of these few threads is not a sure harbinger of building. The spider may spend the whole day in the corner of the cage or hanging from one of the threads. A surer sign of progress to building is indicated when, as in Fig. 34, roughly

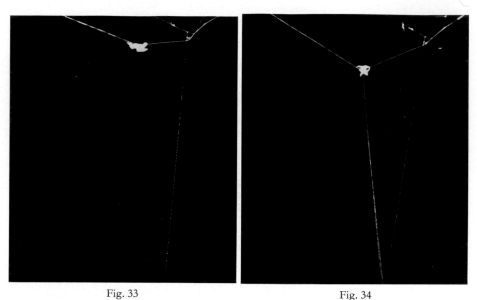

Fig. 33   Fig. 34

Fig. 33. Early stage in building. Several long lines with spider prolonging horizontal thread

Fig. 34. Spider pausing at juncture of threads which becomes the hub. Note that one central angle has already been closed

opposing lines have been drawn together in the upper half of the building space. The fittings and adjustments which now occur are complex. Threads are shifted, tightened, extended, even removed. Extension occurs by the process of drawing in one thread as a new thread is pulled from the spinnerets. The thread to be extended (or replaced) supports the spider at the front end while the new thread provides support from its attachment point. The spider's body forms a moving link between the threads, the slack portion of the old being gathered and drawn into the mouth.

Events become less ad hoc and somewhat easier to follow once a central hub has been established. This juncture of threads becomes the staging point for all operations, the site at which the stage of progress is assessed by the spider. It is not simply that the hub lies along the route of easiest access from one part of the web to another, but that information is gathered here concerning the state of the structure.

The placement of radial threads does not wait upon completion of the frame: these two elements are part of the same phase of building. SAVORY and others have already pointed out that a frame and radial thread may be constructed in the same operation.

It is correct to say, however, that construction of the frame is completed sometime early in the course of radius-building.

Two forms of radii are to be found in most webs, the simple radius and the Y-structure. The construction of the simple form will be described first.

In Fig. 35, the spider can be seen running along an existing radius at position 3 o'clock. The web is distorted by the spider's weight and movement. As the animal runs along the exit-thread, a dragline is pulled from the spinnerets. When the frame

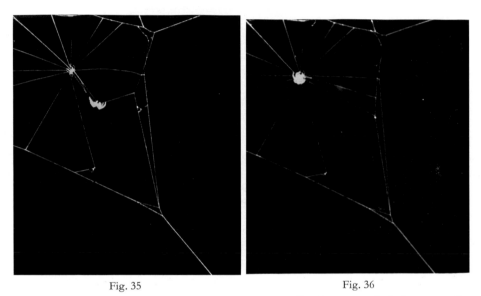

Fig. 35   Fig. 36

Fig. 35. Spider moving from hub to frame along existing radius, in process of placing new radius. Dragline between hub and spinneret is distinct from exit radius. Note, that web is distorted by weight of moving spider, as indicated by slack radii

Fig. 36. New radius in position; spider has returned to hub

is reached, the dragline is attached a short distance below the exit thread, and the spider appears to return along the dragline to the hub (Fig. 36). So rapidly does this action occur that it would seem that the thread must be doubled, as it clearly is in some instances, but usually the dragline is cut and gathered in as the spider makes its way to the hub, in the manner described above. The new radius is usually in place in a matter of 3 to 5 seconds. There is no hesitation at the frame, no grasping or groping of the sort which characterizes the spider once it has returned to the hub.

The second type of radius is illustrated in Fig. 37. Here the animal exited from the hub by way of the radius at position 8 o'clock, moved down the frame thread, attached the dragline to the frame, and began, as before, to return to the hub along the newly-placed line. However, it did not proceed directly to the hub, but attached a line a short distance along the line from the frame, dropped to the frame and pulled the new structure taut into a Y-shape: in effect, a doubly-attached radius. The spider returned to the hub along the arm of the Y; the base of the Y was removed and simultaneously replaced in the same manner as in the simple radius.

The position of the new radius depends upon the thread which has been used to exit from the hub. Fig. 38 shows the same sector of the web being filled on two different occasions. In 38a, the exit-thread was the line in the five o'clock position; the spider can be seen starting its descent. In 38b, the new radius has been placed just clockwise to the exit-radius. The new placement was then burned away with a hot wire so that the sector was returned to its unfilled state. On a subsequent turn about the hub, the spider again encountered this open sector, but this time left the

Fig. 37. Y-structure radius. Exit thread is just above the Y, between 8 and 9 o'clock positions

hub via the radius at position 7 o'clock and placed the new radius just counterclockwise to the exit-radius (Fig. 38c, d). These illustrations are taken from an experiment in which radial placements were repeatedly removed by the author; the flocculent masses on the framing threads are the residues of previous threads which were burned through. In these experiments, it was found that threads tended to be replaced in the same position along the framing thread, with relatively few exceptions. Sometimes, after repeated removals and renewals, the attachment point was moved slightly; on the next replacement, however, it might be returned to the original site. Similarly, repeated removals occasionally provoked a resort to a Y-structure in place of a simple radius, as if a double-mooring were contrived to replace the unreliable single attachment (Reed, 1967).

The Y-structures tend to be associated with mooring positions, i.e. with sectors at convexities of the frame.

The placement of radial threads, simple and Y, proceeds about the web, the spider returning to the hub after each placement. There it circles, its legs successively touching each strand in the process. The new radii tend to be placed below the exit-radii: the spider, hanging from the exit-radius as it scrambles along, encounters the frame thread below the exit thread and grasps it.

There is no simple pattern of placement evident in the evolving web. Various investigators, especially Hingston (1927), Koenig (1951), Peters (1947), have recorded the sequence of placement of radial threads or have experimented with the web in the process of construction. PETERS has sought correspondence between the dimensions or form of the web and the dimensions or form of the spider's body

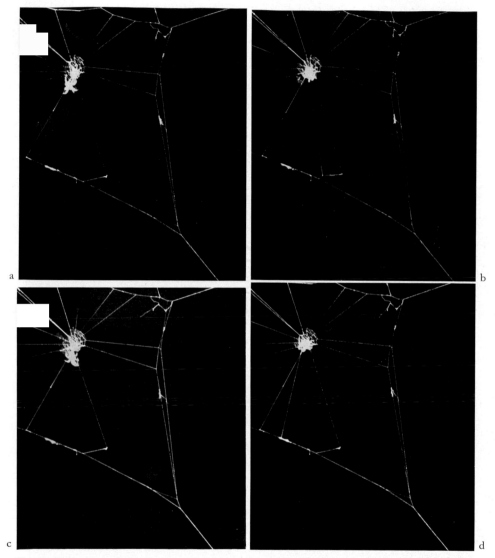

Fig. 38. This series of pictures illustrates the relationship between exit-thread and position of new thread. Spider exits at a) 5 o'clock and c) 7 o'clock and consequently sets new radii at b) between 5 and 6 and d) between 6 and 7

(1947). A common interpretation is that the animal responds to the size of the central angle formed by two adjacent radii (KOENIG; HINGSTON). HINGSTON's belief that the spider moved a certain number of steps from exit-thread to place the new radius is not verified in *Araneus*. TILQUIN's conception of the web as an emergent from a

6 Witt, A Spider's Web

dynamic field of forces (1942) is useful, but he was unable to specify the nature of the attractions and repulsions which he postulated to enter into the process. Speculations such as these, which have sought explanation in response to the tension-state of the web, are undoubtedly fruitful ideas, but they have not yet been accompanied by the kind of mechanical analysis of the process which seems required for elucidation of the essential forces and the spider's means of detecting them. Tracing radial placements thread by thread may verify the validity of such an analysis, but the effort has not yielded insight into the forces. The observation that threads tend to be built 180 degrees to each other would have been instructive if it were invariably true, but it is not.

In fact, at any given stage of radial construction, several open sectors are candidates for filling by the spider, and will be filled if the spider encounters them. Early in the building, there may be four or five regions equally effective in providing triggering cues to radius-spinning. This assertion can easily be tested by removing radial threads when the spider has reached any stage of construction with the possible exception of the final thread (REED, 1967). If, for instance, the most-recently placed radius and all that may be built by the spider subsequently, are carefully burned away as soon as possible after each is built, it will be seen that a set of renewals particular to that stage of construction will be made. Some radii are replaced more frequently than others; they tend to be those which occupy large sectors of the web, but there are enough exceptions to this principle to suspect that size of the central angle is not a sufficient cue to action. Large central angles are tolerated in upper portions of the web; sectors of equal or larger size are filled in the lower sectors. One significant factor in frequency of renewal for a given sector may be the spider's manner of moving on thread. It must suspend itself from the thread by means of its tarsal claws. On returning to the hub, it is more likely to encounter threads below that on which it is moving than it is to touch higher threads. Hence inventory is resumed with these lower sectors. In any event, it is possible that the relative frequency of renewal depends upon the manner of the spider's return to the hub rather than upon some compelling feature of the open sector.

Nevertheless, because of the repeated renewal of the same threads in the candidate-sectors, it is important to examine these for common features. The activity of the spider at the hub suggests that it is there that the essential cues are obtained. The animal appears to actively elicit information, circling and grasping each thread in turn. Coming to an open sector, it may pause briefly and spread both front legs across the radii forming the sector, resting its weight upon them or tugging slightly before proceeding to build or to pass on.

It is not clear what the nature of the testing is. It may simply be a threshold for angular extent which, if exceeded, triggers radius-activity. Alternatively, the cue utilized by the spider might lie in the tension state of the web; here there are a number of possibilities. The spider tugs on the strands; resistance to those tugs stimulates receptors in the legs. Magnitude of stimulation could increase with the angular separation of the threads. This mechanism would then be a kind of response to angle size but on the basis of response to pull rather than on the basis of simple positioning of the legs. Receptors might be symmetrically stimulated according to the vectors of force. SAVORY has hinted at such a possibility. Secondly, the distance to the attachment of each radius may change the magnitude of response to tugging.

Thirdly, placement may depend upon an array of forces coming from several radii simultaneously: not only the position of the first legs, but the position of the other legs and the forces exerted by the weight of the spider's body, would be required for a sufficient analysis of the field of forces.

The spider is sensitive to minute changes in the tension state of the web (HOLZ-APFEL, 1933). If the leg happens to be on a thread which is gently parted by a hot

Fig. 39

Fig. 39 and 40. Close view of hubs of webs of *Zygiella-x-notata*, showing thread detail of central portions of radii and small bridging threads

wire, the animal startles and halts work, but it also does so under less obvious circumstances. In prey capture, vibrations produced by the wings and the struggle of the prey alert the spider, but do not appear to be sufficient to allow precise location of prey. It is necessary to greatly slow or stop motion picture frames to observe that the spider tugs at the strands to locate the prey. It is true that this action may also serve the function of assessing the nature of the contents of the web. One sign of disturbance of the spider — occurring for instance when a cage is moved too roughly — is a violent simultaneous contraction of all legs, so that the animal shakes perpendicular to the plane of the web. This action could serve to reveal the contents of the web. On the other hand, there are other services it might perform: the move-

6*

ment would serve to enmesh prey in the threads, dissuade or make difficult the attack of a predator. There is no reason to limit the adaptive consequences of the action to a single function.

This sensitivity which is inferred from the behavior of the animal contrasts with the orb-weaver's primitive visual capacities. These faculties may also be assessed behaviorally, without recourse to the anatomical evidence of limited visual apparatus.

Fig. 40

The blinded orb-weaver or one building in a light-tight box is not handicapped in its construction of the web. A fly will pass before the spider without hindrance if mechanoreceptors are not somehow stimulated (BALTZER, 1923).

Ordinarily, the process of radius-building continues until the spider makes a complete circuit of the hub without detecting the necessity for additions, however that necessity is signalled. The first complete circuit reveals a very orderly thread which is attached across each of the central angles, apparently placed as the spider moved from radius to radius (Figs. 39 and 40). It seems likely that this tiny bridge was being made all along, adding to the steady accumulation of material at the hub. At any rate, the first complete turn shows it clearly, the second moves the spider out a small distance from the first, and the third and fourth appear in a spiral path out to the edge of the web.

The passage into spiral construction from radial construction has an imperative of its own. As KOENIG has shown, the spider will not endlessly renew radii. There is great variation in the number of replacements which will be made, perhaps depending upon a changing tolerance for incompleteness as the structure nears completion. KOENIG's finding that *Araneus* replaces something like three times the number of threads being used as radii in its webs around the time of the experiment suggests a fixed supply. New material is rather rapidly synthesized in the glands. If the spider pauses in construction — as it may do for several hours if alarmed — substantial replenishing of material could occur. The hypothesis that the timing of phases depends upon the state of the thread glands is supported by PEAKALL's discovery of what appears to be a receptor in the gland (see Chapter 2).

The spiral which terminates radius-building is a provisional structure: it is replaced during the construction of the viscid spiral. It appears to provide support and guidance during the construction of the permanent spiral, though HINGSTON has reported successful completion of the viscid spiral even with the temporary removed.

The transition from the provisional to the permanent spiral is marked by a definite pause in operations. The spider halts for a few seconds; when it again goes into activity, it usually, but not invariably, reverses the direction in which it had been travelling when making the provisional structure. This pause, it seems reasonable to believe, is dictated by a kind of gear-shifting. The material from the aggregate glands is now being included in the line, and some time appears to be required to bring the glandular sources into joint operation.

The spider now works back toward the hub, attaching viscid thread to each radius as it goes. Each turn of the spiral is placed just inside the previous turn. Only toward the center of the web does the spiral become continuous. Usually direction is reversed many times, especially in the lower portions of the web. As Fig. 41 shows, the spiral threads are parallel, even at the reversal points. Not one, but two attachments are made when a reversal occurs, one terminating the original path, the other inaugurating the path in the opposite direction. This practice gives the spiral a discontinuous appearance, but the steady route of the spider inwards can be retraced if this maneuver is borne in mind.

Leg function in the spiral placement serves both locomotion and guidance. The first legs especially seem to be utilized for probing (PETERS, 1932; JACOBI-KLEEMAN, 1953). In this phase of building, the legs may be observed in rapid, rushing motions; they are recorded as blurred images in Fig. 42. The extended leg vibrates in four or five pulsations, each of which bring the limb to a slightly more extended position until contact is made with the existing viscid spiral, or as illustrated in the figure, the provisional spiral.

The outer rear leg pulls thread from the spinnerets as the body of the spider passes from attachment point on one radius to attachment point on another. To secure the threads, the abdomen is brought into contact with the radius. At that moment, the inner rear leg stretches the thread close to the attachment point. The leg is thrust through the plane of the web; the thread is released as the leg is extended. This action appears to contribute to the dispersal of tacky substance along the thread (SAVORY, 1951, p. 88). Under the microscope, the viscid material is seen arrayed in drops along the thread, an arrangement which probably increases its adherent

Fig. 41. Characteristically asymmetric web of *Araneus diadematus* with pendulum turns at bottom and right, and continuous spiral toward center

Fig. 42. Spider placing permanent spiral. Thread is pulled from spinneret with left hind leg while right front leg probes with vibratory motion

characteristics (LANGER, 1967). The apparently greater thickness of the permanent spiral as it appears in photographs is an artifact of the diffraction produced by these droplets. The movement of the spider in this phase of building is economic, rhythmic and relatively difficult to disturb. There appear to be few surplus movements; each

Fig. 43. Permanent spiral construction in progress. At different points in progress of building radii have been cut just central to the tacky spiral. Spider has laid thread across the wide central angle so framed, creating distinct triangular holes in the catching area

leg repeats the same task over and over again. An observer may stand close, even bring a magnifying glass near the spider — a procedure which is likely to terminate its activity in any of the previous stages. For all of its efficiency, spiral construction occupies the greatest amount of the time involved in web-building. For an adult female *Araneus*, radial and structure elements can be accomplished well under 5 minutes of time; the construction of the viscid spiral takes another 20 minutes or longer.

The process also tends to be resistant to disruption of the structural threads. If the radius is destroyed between the provisional spiral turn and the viscid spiral, the spider usually bridges the gap, attaching the thread at the usual interval from the previous spiral turn where it attaches to the radii. The removal of thread produces some contraction of that sector toward the periphery; the new thread forms the base chord of a wedge-shaped gap in the catching zone. Successive wedges can be produced by continuing to remove threads (Fig. 43). If this removal becomes extensive enough, the spider will return to the sagging region of the web, traversing it and pulling the strand into a semblance of tightness, but rather a ragged one. This repairing action, however reluctantly engaged, indicates that the spider pos-

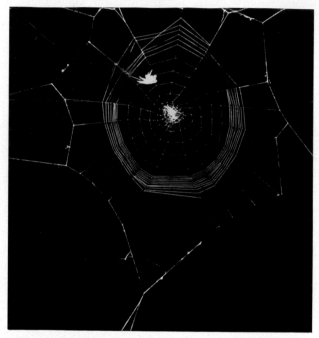

Fig. 44. Spider has laid a number of sticky spiral turns, moving from the web periphery toward the center; the remnants of the temporary spiral are removed as encountered. (Photograph is from unsprayed web — not all parts of structure are illuminated)

sesses both a capacity to allow some degree of disorder and a threshold for reparative action.

Similarly, the building of the provisional spiral is not usually interrupted once begun, although occasionally a radius is belatedly placed. PETERS has filmed the repair of a mooring thread which he had burned away after the permanent spiral was already completed.

The provisional spiral is removed by the spider, bit by bit, as it is encountered during the building of the permanent spiral. Fig. 44 shows the temporary spiral being removed. Small fragments of material may remain at the site of the provisional spiral and, while it is possible that the spider simply detaches the thread, it is more likely that the material is also eaten. Precedent can be found at earlier and later stages of building. Accumulations of thread can be observed to be sucked up during radius-building. Balls of thread-material, rolled up before the spider, suddenly disappear from view, sucked in by the spider as it pauses at the hub. And in the final events of building — when the spider has halted the permanent spiral short of the hub and left a "free zone" between hub and spiral — the flocculent tangle left by operations at the hub is cut out and devoured. It may be that the thread-material has to compose a minimum mass in order to be taken up by sucking, but in the absence of evidence on this point, it seems just as likely that small bits of thread removed in the course of building may also be ingested.

Fig. 45. Adult female *Araneus diadematus*, dorsal view. The animal is seen in the upside down position, with the legs spread apart, in which it awaits contact of prey with web. Note the white cross at the frontal end of abdomen, which identifies the species

The final adjustments at the hub consist of the removal of the tangled remains of the earlier work of radius-building. The holes exposed in the platform may be left open or may be criss-crossed by new threads. The finished hub is a platform upon which the radii converge — a plexus of strands rather than a single point. There the spider takes its station, abdomen upward, legs extended to grasp radial threads, vigilant for the first contact of insect with web (Fig. 45).

C. The Neurophysiological Basis of Web-Building

The neurophysiological events accompanying these intricate behaviors await study. Recent studies of Araneid central nervous system anatomy by BABU (1965) and MEIER (1967) have supplemented the fundamental work of HANSTRÖM, but structure and function are presently related by rather longspanning correlations such as the presence or size of ganglia with relative complexity of behavior. Web-building, a complex behavior, should presumably be richly represented in the central nervous system of Argiopidae, but this requirement does not necessarily imply localization. Web-building is not likely to be a matter of the operation of a sub-system but of the whole central nervous system.

Be that as it may, the central nervous system of the spider conforms to the general Arthropod plan, with departures related to the ecological and behavioral niches

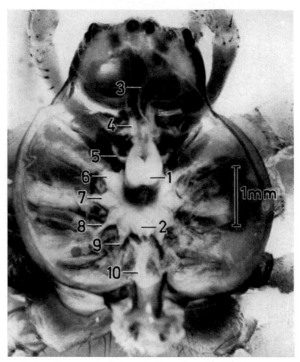

Fig. 46. Dorsal view of central nervous system of *Araneus diadematus* Cl. Carapace has been removed to expose supraesophageal ganglion (1) and sub- or infraesophageal ganglion (2). The ocular nerves (3) and left cheliceral nerves (4) emerge from the rostral end of the supraesophageal ganglion, and the left pedipalp nerve (5) from the subesophageal mass. Numbers 6 through 9 designate the nerves to the first through fourth legs, respectively. Abdominal nerve cord (10) occupies ventral position, running through pedicel into abdomen. Abdomen is removed from this preparation. (Preparation and photograph, courtesy of L. le Guelte, Raleigh)

which the spiders occupy. Designation of structures is largely derived from the study of insects, and the functional implication of the designations must be transferred with caution.

The system is a fused mass in the prosoma; it is a complex neuropile surrounded by cortex containing cell-bodies. It is conventionally divided into a supraesophageal ganglion, or brain (in the words of Bullock and Horridge, the "seat of the long-term organized behavior patterns", 1965, p. 816) and the ventral cord including subesophageal ganglia, serving mouth parts, pedipalps and legs. The gross anatomy of the system is illustrated in Figs. 46 and 47.

A portion of the arthropod brain designated the protocerebrum contains structures receiving input from frontal organs and eyes. A neuropile mass in the protocerebrum, the central body, lies in a position posterior to that found in insects. Usually it is the highest integrating center for visual information, but such function cannot require extensive structure in the web-building spider. Its size, position and dense fiber structure bespeak a coordinating center for the orb-weaver (MEIER, 1967, p. 85); it possesses clear afferent and efferent connections with ganglia of the legs and with the cerebral ganglion, an undifferentiated fiber mass homologous to the deuto-cerebrum. While the size of the central body cannot be related to behavioral com-

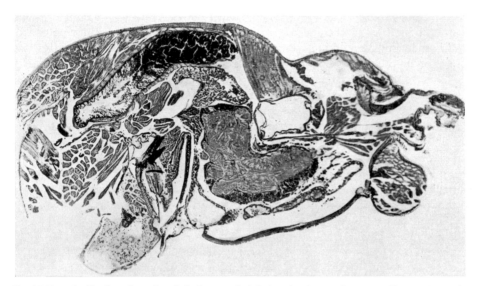

Fig. 47. Longitudinal section of cephalothorax of adult female *Araneus diadematus* Cl., anterior end at left. Boot-shaped structure close to the center of the picture is the central nervous system. Because the section is not at midline, the esophagus does not separate the supra from the infra-esophageal ganglion. Note dark cell cortex enclosing complexly organized neuropile (light gray). The central body forms a distinct round mass at upper right of supraesophageal ganglion

plexity in arthropods generally, it is large in the web-spinning Araneids, smaller in spiders of other habits (BULLOCK and HORRIDGE, 1965, p. 816). Another arthropod feature, neuropile masses called corpora pedunculata — also optic integration centers — are correlated with behavioral complexity, except in the spiders. They are rudimentary or lacking in the orb-weavers.

MEIER found the cerebral ganglion to be richly supplied by fibers from the central body and other parts of the protocerebrum, but with no afferent or efferent connections with the peripheral nervous system, although it supplied efferents to all parts of the central nervous system. He therefore proposed that this undifferentiated neuropile mass composed a coordination center, with the processing of information not restricted to the central body (MEIER, 1967, p. 76).

The ventral ganglia, serving the legs, were found by MEIER to have considerable interconnection ipsilaterally and contralaterally, a finding which seems to reflect coordinative possibilities at lower than the brain levels.

It is, of course a far cry from the evidence of the possibilities for sensory and motor coordination to the synthesis of information into the elucidation of web-building.

As an essential beginning on that synthesis, it seems important to discover what sensory cues have to be integrated by the animal during the behavior. Lesions in the protocerebral region of the brain disturb the precision of construction of the web, as the data reported by WITT (1967) show, but do not make clear the precise nature of the interference which is being produced.

Mechanoreception is the class of sensory input which is most likely to be involved in thread-placement. Photoreception contributes perhaps not at all, although MEIER found the optical ganglia less primitive than previously believed to be the case on HANSTRÖM's evidence.

Several kinds of mechanical cues are conceivably involved: tactile, vibratory, or strain-producing. In addition, various proprioceptive mechanisms may provide information on the position of legs with respect to the body.

We have seen the tactile requirement enter rather clearly into the laying of the viscid spiral, the spider apparently being guided by the next outer (and older) turn of the spiral. MEIER, who has proposed specialization within the order of spiders according to development of central nervous system structures, has classified Argiopidae as a family with well-developed tactile sense. If tactile stimulation refers to the deformation of sensory hairs or trichobothria, presence or absence of a stimulating body would be indicated. The receptors would have to be situated at joints, signalling limb position by number of hairs stimulated or by the particular position of the receptors, if more than presence and absence were to be coded.

COHEN has described how position of limbs is detected in the case of the crab *Cancer magister* Dana (COHEN, 1963). A receptor associated with muscle — the myochordotonal organ — in the merocarpopodite joint was shown, by means of electrophysiological recording, to signal movement and position of the limb. The utility of such an arrangement for the orb-weaver gauging separation of threads is obvious. The small leg-nerve described by PARRY in spiders (PARRY, 1960) which is mainly sensory, seems a good candidate for providing the necessary proprioceptive function.

If the spider responds to separation of threads at the central angle formed by the radii, for instance, some positional receptor system is required. Position, however, may refer to the limb with respect to the body or the limbs with respect to each other, e.g. the first pair of legs.

But the spider may respond not to separation of threads but to states of tension in them. If that is the case, the significant stimulus features are to be found in the geometric disposition of a number of threads and in the resultant forces acting on the spider's body when force is applied to the threads. There are occasions when the spider may be clearly observed to employ a tugging or plucking motion. For instance, while vibration produced by the insect in the web first alerts the spider, the threads are plucked several times rapidly before the exit from the hub is made. The resistance produced by the prey's weight is gross with respect to the resistance to be found in tugging empty radial threads, to be sure, but a subtly sensitive mechanoreceptor has already been described which may provide the necessary cues in building.

The lyriform organs, first described at the close of the last century and variously assigned mechanical and chemical reception functions, were shown by PRINGLE to be stimulated by deformation of the cuticle (PRINGLE, 1955). Briefly, the lyriform organ consists of a number of slits, each of which is served by a single neuron. Opening of the slit produces stretch in dendrites attached to a dome of tissue forming the basement of the slit. WALCOTT has found the lyriform organ to be delicately receptive to vibration (WALCOTT and VAN DER KLOOT, 1959; WALCOTT, 1967) but has not been able to relate this sensivity to the role of the receptor. It may be that a vibratory stimulus provokes response in a receptor not primarily employed for the reception of such stimuli; the significant matter for the problem at hand may be the sensitivity to small displacements of legs which WALCOTT's findings would seem to indicate, and which would be well employed in testing the tension state of threads of the web.

If the building of the web is a reflection of the response of the spider to the tension state of an assemblage of threads, the geometry of the web at various stages of building could provide evidence on the magnitudes to which the hypothetical receptor system is responsive. The situation is complex but soluble. An assemblage of five threads differs in tension dynamics from an assemblage of 16 threads or, for that matter, from an assemblage of five threads which happen to be arrayed differently in space, more or less remotely moored to solid objects. By destroying radii as they are built, *Araneus* can be forced to show which sectors of the web are equivalent as far as requiring bracing at a particular stage of building (REED, 1967). Size of the central angle is not a sufficient characterization of the sector which repeatedly elicits replacement by the spider. A related feature, the tension state of the existing structure, may provide the adequate cue.

This brief review of nervous elements and receptors shows that we are far from being able to integrate the behavioral record with anatomical and physiological data. The event of orb-weaving sets clear problems, the chief one at the moment being the sensory guidance to the behavior. The completed web is an intricate geometric product, but the spider is not a geometer. It is an animal of simple capacities presumably responding to a simple feature of the web, executing a program which cannot be reeled off without some kind of contemporary sensory input. The simple feature and the simple animal yet remain to be elucidated.

VI. References and Acknowledgements

This list of references should help the reader to continue his study within the domain of subjects we have covered. Because our attention has been restricted, many of the familiar topics of arachnology have been slighted, notably systematics, mating behavior and venoms. The wonderful books for children on spiders and web-building have been omitted. Books and papers, including some often quoted in the literature, were left from the list if they could not be found, on the assumption that the reader would experience similar difficulty. We have pruned away short communications if they were discussed in extenso in later papers of the same author, although otherwise we have omitted discussions or accounts which were merely reiterative. Despite these exclusions, there are more papers in the list than are quoted in the text; full titles will assist the judgement of the relevance of a paper to a particular interest.

Laboratory investigation of the physiology and behavior of spiders requires patient and reliable assistance. Many valued helpers are mentioned in the text, but none surpass the work of Mrs. MARGUERITE AGI, who spent more than 8 years collecting and analyzing data, handling the animals and developing method.

Financial support for the work reported here came first from Sandoz Pharmaceuticals in Basel, later from the Swiss National Fund, the National Institutes of Health and, most generously, from the National Science Foundation. Some aspects of the research were assisted by grants from the American Heart Association. To all these agencies go our thanks; we hope that some reward for their expenditures is to be found in these pages.

AKAI, H., and M. KOBAYASHI: Sites of fibroin formation in the silk gland in *Bombyx mori*. Nature (Lond.) **206**, 529 (1965).

ALBIN, E.: A natural history of spiders, and other curious insects. London: John Tilly 1736.

AMBROSE, E. J., C. H. BAMFORD, A. ELLIOTT, and W. E. HANBY: Water-soluble silk: An α-protein. Nature (Lond.) **167**, 264—265 (1951).

APSTEIN, C.: Bau und Function der Spinndrüsen der Araneida. Arch. Naturg. **55**, 29—74 (1889).

ASTBURY, W. T.: Problems in the x-ray analysis of the structure of animal hairs and other protein fibres. Trans. Faraday Soc. **29**, 193—211 (1933).

—, and A. STREET: X-ray studies of the structure of hair, wool and related fibres. 1. General. Phil. Trans. R. Soc. A **230**, 75—101 (1931).

BABU, K. S.: Certain histological and anatomical features of the central nervous system of a large Indian spider *Poecilotheria*. Web Building Spiders. AAAS Section on Zoology, New York, 1967 (in preparation, Amer. Zoologist, 1968).

— Anatomy of the central nervous system of arachnids. Zool. Jb. Abt. Anat. **82**, 1—154 (1965).

BACQ, A. M.: Metabolism of adrenalin. Pharmacol. Rev. **1**, 1—26 (1949).

BALTZER, F.: Beiträge zur Sinnesphysiologie und Psychologie der Webespinnen. Mitteilungen der Naturforschenden Gesellschaft in Bern 163—187 (1923).

BALTZER, F.: Über die Orientierung der Trichterspinne *Agalena labyrinthica* (Cl.) nach der Spannung des Netzes. Rev. suisse Zool. **37**, 363—369 (1930).

BAMFORD, C. H., A. ELLIOTT, and W. E. HANBY: Fibrous proteins and their relation to synthetic polypeptides. In Synthetic Polypeptides. Preparation, Structure and Properties. Acad. Press, N.Y. **12**, 369—408 (1956).

BARROWS, W. M.: The reaction of an orb-weaving spider. *Epeira sclopetaria* Cl., to rhythmic vibrations of its web. Biological Bulletin **29**, 316—332 (1915).

BAYS, S. M.: A study of the training possibilities of *Araneus diadematus* Cl. Experientia (Basel) **18**, 423 (1962).

— A mathematical model approach to radius construction. Web Building Spiders. AAAS Section on Zoology, New York, 1967.

BELL, A. L., and D. B. PEAKALL: Changes in fine structure during silk protein production in the ampullate gland of the spider *(Araneus sericatus)*. J. Cell Biol. (in press).

BERCEL, N. A.: A study of the influence of schizophrenic serum on the behavior of the spider *Zilla-x-notata*. Arch. Gen. Psychiatry **2**, 189—209 (1960).

BERLAND, L.: Les Arachnides. Encyclopedie Entomologique. Paris: Lechevalier 1932.

BLUM, M. S., J. P. WOODRING, and E. N. LAMBREMONT: Qualitative and quantitative analysis of the fatty acids in the spider *Nephila clavipes* (Araneida: Argiopidae). Ann. Entomol. Soc. Am. **56**, 345—348 (1963).

BOEUFFLE, R. L.: L'épeire et sa toile. Bull. Soc. Geol. Norm. Mus. Havre **35**, 133—147 (1931).

BONNET, P.: Bibliographia araneorum. Toulouse **1**, 3 (1945).

— Nouvelles recherches sur les griffes des pattes des araignees. Bull. Soc. Hist. Nat., Toulouse **67**, 346—352 (1935).

— Thèses. La mue, l'autotomie et la régénération chez les araignées, avec une étude des dolomèdes d'Europe. Imprimerie Toulousaine, 1930.

BOYS, C. V.: The influence of a tuning-fork on the garden spider. Nature (Lond.) **23**, 149—150 (1880).

BRADFIELD, J. R. G.: Phosphatases and nucleic acids in silk glands: Cytochemical aspects of fibrillar protein secretion. Quart. J. Microsc. Sci. **92**, 87—112 (1950).

BRAUNITZER, G., u. D. WOLFF: Vergleichende chemische Untersuchungen über die Fibroine von *Bombyx mori* und *Nephila madagascariensis*. Z. Naturforsch. **10** b, 404—408 (1955).

BREED, A. L., V. D. LEVINE, D. B. PEAKALL, and P. N. WITT: The fate of the intact orb web of the spider *Araneus diadematus* Cl. Behaviour **23**, 43—60 (1964).

BRISTOWE, W. S.: A book of spiders. London-New York: The King Penguin Books 1947.

BUCHLI, H.: Notes préliminaires concernant le comportement de chasse et le rhythme d'activité de la Mygale maconne, *Nemesia caementaria* Latreille (1798). Rev. Ecol. Biol. Sol. T. **2**, 403—438 (1965).

— On the hunting behavior of *Ctenizidae*. Web Building Spiders, AAAS Section on Zoology, New York, 1967 (in preparation, Amer. Zoologist, 1968).

BUECHERL, W.: Instintos maternais nas aranhas Brasileiras. Dusenis **3**, 57—74 (1952a).

— Brutfürsorge und Brutpflege bei einigen brasilianischen Spinnen. 9th Inter. Cong. Entomol. **1**, 1091—1092 (1952b).

— Histologia das glandulas de veneno de algumas aranchas e escorpioes. Mem. Inst. Butantan **31**, 77—84 (1964).

— Biology and venoms of the most important South American spiders of the genera *Rhoneutria*, *Loxosceles* and *Latrodectus*. Web Building Spiders, AAAS Section on Zoology, New York, 1967 (in preparation, Amer. Zoologist, 1968).

BULLOCK, T. H., and G. A. HORRIDGE: Arthropoda, Arachnida. Structure and Function in the Nervous System of Invertebrates. San Francisco: W. H. Freeman 1965.

CAZIER, M. A., and M. A. MORTENSON: Analysis of the habitat, web design, cocoon and egg sacs of the tube weaving spider *(Diguetia canities* (McCook) (Aranea, Diguetidae)). Soc. Calif. Academy Sci. **61**, 65—88 (1962).

CHICKERING, A. M.: Evolution in spiders. Rep. Mich. Acad. Sci. **38**, 22—51 (1935).

CHRISTIANSEN, A., R. BAUM, and P. N. WITT: Changes in spider webs brought about by mescaline, psilocybin and an increase in body weight. J. Pharmac. exp. Ther. **136**, 31—37 (1962).

CHRYSANTHUS, F.: Hearing and stridulation in spiders. Tijdschrift voor entomologie **96**, 57—83 (1953).

CLOUDSLEY-THOMPSON, J. L.: Spiders, scorpions, centipedes and mites. New York: Pergamon Press 1958.

COHEN, M. J.: The crustacean myochordotonal organ as a proprioceptive system. Comp. Biochem. Physiol. **8**, 223—243 (1963).

COHEN, N. R.: The control of protein biosynthesis. Biol. Rev. **41**, 503—560 (1966).

COMSTOCK, J. H.: The spider book. Ithaca, N.Y.: Comstock Publishing Co., Inc., 1948.

CRICK, F. H. C.: On protein synthesis, p. 138—163. In: The Biological Replication of Macromolecules. Symp. No. 12, Soc. Exp. Biol. New York: Cambridge University Press 1958.

CROMPTON, J.: The life of the spider. Boston: Houghton Mifflin Co. 1951.

DAHL, F.: Versuch einer Darstellung der psychischen Vorgänge in den Spinnen I. Vierteljahrsschrift für wissenschaftliche Philosophie, S. 84—103 (1885a).

— Versuch einer Darstellung der psychischen Vorgänge in den Spinnen II. Vierteljahrsschrift für wissenschaftliche Philosophie, S. 162—190 (1885b).

DAWYDOFF, C.: Developpement embryonnaire des Arachnides. Traité de Zool. **6**, 320—385 (1949).

DEVOE, R. D.: Linear superposition of retinal action potentials to predict electrical flicker responses from the eye of the wolf spider. *Lycosa baltimoriana* (Keyserling). J. gen. Physiol. **46**, 75—96 (1962).

DOBB, M. G., R. D. B. FRASER, and T. P. MACRAE: The fine structure of silk fibroin. J. Cell Biol. **32**, 289—295 (1967).

DREES, O.: Untersuchungen über die angeborenen Verhaltensweisen bei Springspinnen (Salticida). Z. Tierpsychol. **9**, 169—207 (1952).

EBERHARD, W.: Computer simulation of orb web building. Web Building Spiders. AAAS Section on Zoology, New York, 1967 (in preparation, Amer. Zoologist, 1968).

EISNER, T.: Insects scales are asset in defense. Natural History **74**, 26—31 (1964).

—, R. ALSOP, and G. ETTERSHANK: Adhesiveness of spider silk. Science **146**, 1058—1062 (1964).

EMERIT, M.: La Trichobothriotaxie et ses variations au cours du dévelopement post-embryonnaire chez l'araignée *Gasteracantha versicolor*. (Walck.) (Argiopidae). Acad. Sci. Paris **258**, 4843—4845 (1964).

EMERTON, J. H.: Common spiders of the United States. Boston: Gin & Co. 1920.

EPELBAUM, F.: Die Wirkung akuter und chronischer Kohlenoxyd- und Kohlendioxyd-Vergiftung auf die Spinne *Zilla-x-notata* Cl. und ihren Netzbau. Arch. int. Pharmacodyn. **106**, 275—293 (1956).

ERSKINE, C. A.: Micromanipulation in control and handling of *Zygiella-x-notata* as an experimental animal. Science **133**, 644—646 (1961).

FABRE, J. H.: Souveniers Entomologiques, 6. série, 9 ed. Études sur l'instinct et les moeurs des insects. Paris: Librairie Delagrave 1923.

— The life of the spider. New York: Blue Ribbon Books 1912.

FISCHEL, W.: Wachstum und Häutung der Spinnen. 1. Mittlg. Studien an retitelen Spinnen. Z. wiss. Zool. **133**, 442—469 (1929).

— Wachstum und Häutung der Spinnen. 2. Mittlg. Weitere Beobachtungen an retitelen und vaganten Araneen. Z. wiss. Zool. **136**, 78—107 (1930).

FISCHER, F. G., u. J. BRANDER: Eine Analyse der Gespinste der Kreuzspinne. Hoppe-Seylers Z. physiol. Chem. **320**, 92—102 (1960).

FITCH, H. S.: Spiders of the University of Kansas Natural History Reservation and Rockefeller Experimental Tract. University of Kansas 1963.

FRANK, H.: Untersuchungen zur funktionellen Anatomie der lokomotorischen Extremitäten von *Zygiella-x-notata*, einer Radnetzspinne. Zool. Jb. Abt. Anatomie **76**, 423—460 (1957/58).

FREISLING, J.: Netz und Netzbauinstinkte bei *Theridium saxatile* Koch. Z. wiss. Zool. **165**, 396—421 (1961).

FRINGS, H., and M. FRINGS: Reactions of orb-weaving spiders (Argiopidae) to airborne sounds. Ecology **47**, 578—588 (1966).

GARDNER, B. T.: Observations on three species of *Phidippus* jumping spiders (Araneae: Salticidae). Psyche **72**, 133—147 (1965).

GERHARDT, U.: Neue biologische Untersuchungen an einheimischen und ausländischen Spinnen. Z. Morph. Ökol. der Tiere **8**, 96—186 (1927).

GERTSCH, W. J.: American spiders. New York: D. van Nostrand Co. 1949.

— The spider genus *Zygiella* in North America (Araneae, Argiopidae). Amer. Mus. Nat. Hist. **2188**, 1—21 (1964).

GOERNER, P.: Mehrfach innervierte Mechanorezeptoren (Trichobothrien) bei Spinnen. Naturwissenschaften **52**, 437 (1965).

GOLDSTEIN, L., and M. PRESCOTT: Protein in nucleocytoplasmic interactions. 1. The fundamental characteristics of the rapidly migrating proteins and the slow turnover proteins of the Amoeba proteus nucleus. J. Cell Biol. **33**, 637—655 (1967).

GROH, G., et M. LEMIEUX: Essai comparative de deux médicaments antipsychotiques sur la formation de la toile d'araignée. Trimipramine, a new antidepressant. Colloquium St. Jean-de-Dieu Hospital Montreal-Gamelin, Quebec, 28. 5. 1964, p. 35—42.

HANSTROEM, B.: Fortgesetzte Untersuchungen über das Araneengehirn. Zool. Jb. Anat. **59**, 455—478 (1935).

HEIMANN, H., u. P. N. WITT: Die Wirkung einer einmaligen Gabe von Largactil auf den Netzbau der Spinne *Zilla-x-notata*. Mschr. Psychiat. Neurol. **129**, 104—128 (1955).

HENTZ, N. M.: Spiders of the United States. Boston: Society of Natural History 1875.

HINGSTON, R. W. G.: A naturalist in Hindustan. Boston: Small, Maynard and Co.

— A naturalist in Himalaya. London: H. F. & G. Willoby 1920.

— Protective devices in spiders' snares, with a description of seven new species of orb-weaving spiders. Proc. Zool. Soc. **18**, 259—293 (1927).

— A naturalist in the Guiana forest. London: Longmans, Green and Co. 1932.

HOFFER, A., H. OSMOND, and J. SMYTHIES: Schizophrenia: a new approach II. Result of a year's research. J. ment. Sci. **100**, 29—45 (1954).

HOKIN, L. E., and M. R. HOKIN: The synthesis and secretion of digestive enzymes by pancreas tissue in vitro. In Ciba Foundation Symposium on the Exocrine Pancreas, p. 186—207. A. V. S. DE REUCK, and M. D. CAMERON, ed. Boston: Little, Brown and Co. 1961.

HOLZAPFEL, M.: Die Bedeutung der Netzstarrheit für die Orientierung der Trichterspinne *Agalena labyrinthica* (Cl.). Rev. suisse Zool. **40**, 247—250 (1933a).

— Die nicht-optische Orientierung der Trichterspinne *Agalena labyrinthica* (Cl.). Z. vergl. Physiol. **20**, 55—115 (1933b).

HOMANN, H.: Beiträge zur Physiologie der Spinnenaugen: IV. Das Sehvermögen der Thomisiden. Z. wiss. Biol., Abt. C. Physiol. **20**, 420 (1934).

HURWITZ, J., J. J. FURTH, M. MALAMY, and M. ALEXANDER: The role of DNA in RNA synthesis. 3. The inhibition of the enzymatic synthesis of RNA and DNA by actinomycin D and proflavin. Proc. nat. Acad. Sci. (Wash.) **48**, 1222—1230 (1962).

JACOBI-KLEEMANN, M.: Über die Lokomotion der Kreuzspinne *Aranea diadema* beim Netzbau (nach Filmanalysen). Z. vergl. Physiol. **34**, 606—654 (1953).

KADZIELA, W., and W. KOKOCINSKI: The effect of some neurohormones on the heart rate of spiders. Experientia (Basel) **22**, 45—46 (1966).

KAJAK, A.: Quantitative analysis of relations between spiders *Araneus cornutus* Clerck and *Araneus quadratus* Clerck and their prey. Bull. Acad. Pol. Sci. Ser. Sci. Biol. **13**, 515—522 (1965).

KARNOVSKY, M. L.: Metabolic basis of phagocytic activity. Physiol. Rev. **42**, 143—168 (1962).

KASTON, B. J.: Spiders of Connecticut. Published by the State of Connecticut 1948.

— How to know the spiders. Dubuque, Iowa: C. Brown Co. 1953.

— The evolution of spider webs. Amer. Zoologist **4**, 191—207 (1964).

— Some little known aspects of spider behavior. The Amer. Midland Naturalist **73**, 336—356 (1965).

— Evolution of the web. Natural History **75**, 26—33 (1966).

KELLER, L. R.: Untersuchungen über den Geruchssinn der Spinnenart *Cupiennius salei* Keyserling. Z. vergl. Physiol. **44**, 576—612 (1961).

KEYSERLING, G. E.: Spinnen Amerikas. Laterigradae, Vol. 1. Bauer & Raspe 1880.

— Spinnen Amerikas. Theridiidae, Vol. 2. Bauer & Raspe 1884.

— Spinnen Amerikas. Theridiidae, Vol. 2. Bauer & Raspe 1886.

—, u. G. MARX: Spinnen Amerikas. Brasilianische Spinnen, Vol. 3. Bauer & Raspe 1891.

— — Spinnen Amerikas. Epeiridae, Vol. 4. Bauer & Raspe 1892.

— — Spinnen Amerikas. Epeiridae. Bauer & Raspe 1893.

KIRCHNER, W.: Wie überwintert die Schilfradspinne *Araneus cornutus*? Natur und Museum **95**, 163—170 (1965).

KOENIG, M.: Beiträge zur Kenntnis des Netzbaus orbiteler Spinnen. Z. Tierpsychol. **8**, 462—493 (1951).

KRAFFT, B.: Sur une possibilité d'échanges de substance entre les individus chez l'araignée sociale
  Agelena consociata Denis. C. R. Acad. Sci. (Paris) **260**, 5376—5378 (1965).
— Premières récherches de laboratoire sur le comportement d'une araignée sociale nouvelle „*Agelena*
  consociata, Denis". Extrait de la Rev. comportement Animal **1**, 25—30 (1966).
— Various aspects of the biology of *Agelena consociata*. Web Building Spiders. AAAS Section on
  Zoology, New York, 1967 (in preparation, Amer. Zoologist, 1968).
KRATKY, O.: X-ray investigation of silk fibroin. Trans. Faraday Soc. **52**, 558—570 (1956).
KUEHNE, H.: Die neurosekretorischen Zellen und der retrocerebrale neuroendokrine Komplex von
  Spinnen (Araneae, Labidognatha). Zool. Jb. Abt. Anat. Ontog. **77**, 527—560 (1958/59).
LANGER, R. M.: The physics of spider silk. Web Building Spiders. AAAS Section on Zoology, New
  York, 1967 (in preparation, Amer. Zoologist, 1968).
—, and W. EBERHARD: Laboratory photography of spider webs. Web Building Spiders. AAAS
  Section on Zoology, New York, 1967 (in preparation, Amer. Zoologist, 1968).
—, and V. FRIEDRICH: Electron microscope observations of spider silk. Web Building Spiders. AAAS
  Section on Zoology, New York, 1967 (in preparation, Amer. Zoologist, 1968).
LÉGENDRE, R.: Thèses — Contribution a l'étude du système nerveux des Aranéides. Paris: Masson
  1959.
— L'audition et l'émission de sons chez les Aranéides. Ann. biol. **7**, 371—390 (1963).
— Morphologie et développement des Chélicerates. Embryologie, developpement et anatomie des
  Aranéides. Fortschr. Zool. **17**, 238—271 (1965).
LE GUELTE, L.: Sur l'élevage et la croissance de l'araignée *Zilla-x-notata* Cl. Bull. Mus. Nat. Hist.
  Naturelle **34**, 380—392 (1962).
— Developpement accelèrè de l'araignée *Zilla-x-notata* Cl. Bull. Mus. Nat. Hist. Naturelle **35**,
  273—274 (1963).
— Instinct et facteurs physiques au cours de la construction de la toile chez *Zilla-x-notata* Cl. Psycho-
  logie Française **9**, 280—286 (1964).
— Repercussions de la perte de pattes sur la construction de la toile chez *Araneus diadematus* et
  Zygiella-x-notata. Psychologie Française **10**, 257—264 (1965a).
— Situation de la rétraite et structure de la toile de *Zygiella-x-notata* Cl. Communication 3. Rencontre
  des Arachnologistes Européens, Mai, 23—25 (Senck. biol.) 1965b.
— Structure de la toile de *Zygiella-x-notata* Cl. et facteurs qui régissent le comportement de l'araignée
  pendant la construction de la toile. Thèse, Publ. Université de Nancy 1966a.
— Note preliminaire sur un apprentissage chez *Zygiella-x-notata* Cl. C. R. Acad. Sci. (Paris) **262**,
  689—691 (1966b).
— Learning in spiders. Web Building Spiders, AAAS Section on Zoology, New York, 1967 (in
  preparation, Amer. Zoologist, 1968).
LENORMANT, H.: Infrared spectra and structure of the proteins of the silk glands. Trans. Faraday
  Soc. **52**, 549—553 (1956).
LIESENFELD, F. J.: Untersuchungen am Netz und über den Erschütterungssinn von *Zygiella-x-notata*
  Cl. (Araneidae). Vergl. Physiol. **38**, 563—592 (1956).
— Über Leistung und Sitz des Erschütterungssinnes von Netzspinnen. Biol. Zentlb. **80**, 465—475
  (1961).
LOEWENSTEIN, W. R.: Biological transducers. Scient. Amer. **203**, 98—108 (1960).
LOUGEE, L. B.: The web of the spider. Cranbrook Institute of Science, 1964.
LUCAS, F.: Spiders and their silks. Discovery **25**, 1—7 (1964).
—, and J. T. B. SHAW: Comparative studies of fibroins. 1. The amino acid composition of various
  fibroins and its significance in relation to their crystal structure and taxonomy. J. molec. Biol. **2**,
  339—349 (1960).
— —, and S. G. SMITH: The chemical constitution of some silk fibroins and its bearing on their
  physical properties. Shirley Inst. Mem. **28**, 77—89 (1955).
— — — The silk fibroins. Advanc. Protein Chem. **13**, 107—242 (1958).
MACKENSEN, O.: Effect of carbon dioxide on initial oviposition of artifically inseminated and virgin
  queen bees. J. Econom. Entomol. **40**, 344—349 (1947).
MARPLES, B. J.: An unusual type of web constructed by a Samoan spider of the family Argiopidae.
  Trans. roy. Soc. N.Z. **77**, 232—233 (1949).
— Notes on spiders of the family Uloboridae. Ann. Zool. **4**, 1—10 (1962).
— The spinnerets and epiandrous glands of spiders. J. Linn. Soc. (Zool.) **46**, 209—222 (1967).

MATIJEVIC, E., R. H. OTTEWILL, and M. KERKER: Light scattering by infinite cylinders. Spider fibers. J. Optic. Soc. Am. **51**, 115—116 (1961).

MAYER, G.: Untersuchungen über Herstellung und Struktur des Radnetzes von *Aranea diadema* and *Zilla-x-notata* mit besonderer Berücksichtigung des Unterschieds von Jugend- und Altersnetzen. Z. Tierpsychol. **9**, 337—362 (1953).

McCOOK, H. C.: The snare of the ray spider *(Epeira radiosa)*, a new form of orb-web. Proc. Acad. Nat. Sci. (Philad.) 163—175 (1881a).

— How orb-weaving spiders make the frame work or foundations of webs. Proc. Acad. Nat. Sci. (Philad.) 430—435 (1881b).

— Snares of orb-weaving spiders. Proc. Acad. Nat. Sci. (Philad.) 254—257 (1882).

— Hibernation and winter habits of spiders. Proc. Acad. Nat. Sci. (Philad.) 102—104 (1885).

— American Spiders and their spinning work, Vol. I—III. Publ. by Author and Acad. Nat. Sci. (Philad.) 1889—1893.

McCRONE, J.: Spider venoms, comparative aspects. Web Building Spiders, AAAS Section on Zoology, New York, 1967 (in preparation, Amer. Zoologist, 1968).

MECQUEM, C. DE: L'araignée épeire diadème et la confection de sa toile. The spider *Aranea diademata* and the construction of its web. Dusser, Larcheveque, published Bourges, 1—32, 1924.

MEIER, F.: Beiträge zur Kenntnis der postembryonalen Entwicklung der Spinnen „Araneida, Labidognatha" unter besonderer Berücksichtigung der Histogenese des Zentralnervensystems. Rev. suisse Zool. Annales de la Societé Suisse de Zoologie. Genève: Kundig 1967.

MENGE, A.: Preußische Spinnen (2 Vols.), Danzig 1866.

MIKULSKA, I.: Changes in heart rate in spiders effected by increased temperature. Zool. pol. **12**, 149—160 (1961a).

— Heartbeat in newly hatched spiders. Katedra Zoologii Systematycznej, Biologia **6**, 21—23 (1961b).

— Parental care in a rare spider *Pellenes nigrociliatus*. (L. Koch) var. *bilunulata* Simon. Nature (Lond.) **190**, 365—366 (1961c).

MILLOT, J.: Contributions a l'histophysiologie des Aranéides. Bull. Fr. belg. Supp. **8**, 1—238 (1926).

— Ordre des Aranéides (Araneae). 1. Morphologie générale et anatomie interne. Traité de Zoologie **6**, 569—743 (1949).

MIURA, Y., H. ITOH, K. SUNAGA, and S. OGOSHI: Studies on the protein synthesis in silkglands. 6. RNA metabolism during fifth instar larvae. J. Biochem. **58**, 293—299 (1965).

MONTGOMERY, T. H., JR.: Studies on the habits of spiders, particularly those of the mating period. Proc. Acad. Nat. Sci. (Philad.) **55**, 59—149 (1903).

NIELSEN, E.: The biology of spiders, Vols. I—II. Copenhagen: Levin and Munksgaard 1932.

PALADE, G. E., P. SIEKEVITZ, and G. CARO: Structure, chemistry, and function of the pancreatic exocrine cell, p. 23—55. In Ciba Foundation Symposium on the Exocrine Pancreas. A.V.S. DE REUCK and M. P. CAMERON, Ed. Boston: Little, Brown and Co. 1961.

PAPI, F., and P. TONGIORGI: Innate and learned components in the astronomical orientation of wolf spiders. Ergebn. Biol. **26**, 259—280 (1963).

PARRY, D. A.: The small leg-nerve of spiders and a probable mechanoreceptor. Quart. J. micr. Sci. **101**, 1—8 (1960a).

— Spider hydraulics. Endeavour **19**, 156—162 (1960b).

— The signal generated by an insect in a spider's web. J. exp. Biol. **43**, 185—192 (1965).

—, and R. H. J. BROWN: The hydraulic mechanism of the spider leg. J. exp. Biol. **36**, 423—433 (1959).

PEAKALL, D. B.: Effects of cholinergic and anticholinergic drugs on the synthesis of silk fibroins in spiders. Comp. Biochem. Physiol. **12**, 465—470 (1964a).

— Composition, function, and glandular origin of the silk fibroins of the spider, *Araneus diadematus* Cl. J. exp. Zool. **156**, 345—350 (1964b).

— Regulation of the synthesis of silk fibroins of spiders at the glandular level. Comp. Biochem. Physiol. **15**, 509—516 (1965a).

— Differences in regulation in the silk glands of the spider. Nature (Lond.) **207**, 102—103 (1965b).

— Regulation of protein production in the silk glands of spiders. Comp. Biochem. Physiol. **19**, 253—258 (1966a).

— Silk synthesis, mechanism and location. Web Building Spiders, AAAS Section on Zoology, New York, 1967a (in preparation, Amer. Zoologist, 1968).

PEAKALL, D. B.: Incorporation of C^{14}-orotic acid and C^{14}-amino acids into pigeon pancreas slices following cholinergic stimulation. Proc. Soc. exp. Biol. (N.Y.) 126, 198—201 (1967b).
— The spider's dilemma. New Scientist 37, 28—29 (1968a).
— Autoradiographic localization of acetylcholine on the ampullate gland of the spider. J. Pharmacol. exp. Ther. 160, 81—90 (1968b).
—, and I. L. CAMERON: Non-conservation of lysine labeled nuclear material in spider silk glands. Exp. Cell Res. 49, 199—202 (1968).
PECKHAM, G. W., and E. G. PECKHAM: Some observations on the mental powers of spiders. J. Morph. 1, 383—419 (1887).
PETERS, H. M.: Über die Orientierung der Insekten und Spinnen. Natur und Museum 62, 318—322 (1932a).
— Experimente über die Orientierung der Kreuzspinne *Epeira diademata* Cl. im Netz. Zool. Jb., Abt. Zool. 51, 239—288 (1932b).
— Weitere Untersuchungen über die Fanghandlung der Kreuzspinne (*Epeira diademata* Cl.). Z. vergl. Physiol. 19, 47—67 (1933).
— Über das Kreuzspinnennetz und seine Probleme. Naturwissenschaften 47, 776—786 (1939a).
— Probleme des Kreuzspinnen-Netzes. Z. Morphol. u. Oekol. d. Tiere 36, 180—266 (1939b).
— Zur Geometrie des Spinnen-Netzes. Z. Naturforsch. 2 b, 227—232 (1947).
— Untersuchungen über die Proportionierung im Spinnen-Netz. Z. Naturforsch. 6 b, 90—107 (1951).
— Zentralnervöse Steuerung bei Araneiden. Experientia (Basel) 9, 183 (1953a).
— Weitere Untersuchungen über den strukturellen Aufbau des Radnetzes der Spinnen. Z. Naturforsch. 8 b, 355—370 (1953b).
— Worauf beruht die Ordnung im Spinnen-Netz? Umschau 12, 268—370 (1954).
— Über den Spinnapparat von *Nephila madagascariensis*. Z. Naturforsch. 10 b, 395—404 (1955).
— Maturation and coordination of web-building activity. Web Building Spiders, AAAS Section on Zoology, New York, 1967 (in preparation, Amer. Zoologist, 1968).
—, u. P. N. WITT: Die Wirkung von Substanzen auf den Netzbau der Spinnen. Experientia (Basel) 5, 161 (1949).
— — u. D. WOLFF: Die Beeinflussung des Netzbaues der Spinnen durch neurotrope Substanzen. Z. vergl. Physiol. 32, 29—44 (1950).
PETRUSEWICZOWA, E.: Beobachtungen über den Bau des Netzes der Kreuzspinne. Travaux de l'Institut de Biologie de l'Université de Wilno 9, 1—25 (1938).
POINTING, P. J.: Some factors influencing the orientation of the spider. (*Frontinella communis* (Hentz) in its web (Araneae: Linyphiidae).) Canad. Entomol. 97, 69—78 (1965).
POLLAK, P. M.: Esterase activity in the Arachnida. Nature (Lond.) 211, 546—547 (1966).
POULSSON, E.: Handbuch der experimentellen Pharmakologie 2, 322 (1923).
PRECHT, H.: Über das angeborene Verhalten von Tieren. Versuche an Springspinnen. Z. Tierpsychol. 9, 207—230 (1952).
PRESCOTT, D. M., and M. A. BENDER: Synthesis and behavior of nuclear proteins during the cell life cycle. J. cell. comp. Physiol. 62 (Suppl. 1), 175—194 (1963).
PRINGLE, J. W. S.: The function of the lyriform organs of arachnids. J. exp. Biol. 32, 270—278 (1955).
QUATREMÈRE DISJONVAL: Naturgeschichte der Spinnen. Frankfurt am Main 1798.
RABAUD, E.: Construction et structure de la toile d'*Argiope Bruennichi* (Construction and structure of the web of *Argiope Bruennichi*). Soc. ent. Fr. Livre du Centenaire 523—535 (1932).
RAMSDEN, W.: Coagulation by shearing and by freezing. Nature (Lond.) 142, 1120—1121 (1938).
RATHMAYER, W.: Neuromuscular transmission in a spider and the effect of calcium. Comp. Biochem. Physiol. 14, 673—687 (1965a).
— Polyneurale Innervation bei Spinnen. Naturwissenschaften 52, 114 (1965b).
REED, C. F.: The way of a weaver. New Scientist 33, 606—608 (1967).
— Cues in the web-building process. Web Building Spiders, AAAS Section on Zoology, New York, 1967 (in preparation, Amer. Zoologist, 1968).
—, and P. N. WITT: Progressive disturbance of spider web geometry by two sedative drugs. Physiology and Behaviour 3, 119—124 (1968).
— —, and R. L. JONES: The measuring function of the first legs of *Araneus diadematus* Cl. Behaviour 25, 98—119 (1965).

REISKIND, J.: Stereotyped burying behavior in Sicarius. Web Building Spiders, AAAS Section on Zoology, New York, 1967 (in preparation, Amer. Zoologist, 1968).

RICHTER, G.: Untersuchungen über Struktur und Funktion der Klebfäden in den Fanggeweben ecribellater Radnetzspinnen. Naturwissenschaften **43**, 23 (1956).

RIEDER, H. P.: Methodische und statistische Beobachtungen zum Spinnentest. Verh. Naturf. Ges. Basel **69**, 4965 (1957a).

— Biologische Toxizitätsbestimmung pathologischer Körperflüssigkeiten: III. Prüfung von Urin-extrakten Geisteskranker mit Hilfe des Spinnentestes. Psychiat. et Neurol. (Basel) **134**, 378—395 (1957b).

— Prüfung von Ausscheidungsprodukten Geisteskranker mit Hilfe des Spinnentestes. Confin. neurol. (Basel) **18**, 226—232 (1958a).

— Über die Wirkung von Histamin, Serotonin und Nor-Adrenalin auf den Netzbau von Spinnen (*Zilla-x-notata* Cl.). Arch. int. Pharmacodyn. **115**, 326—331 (1958).

ROBINSON, M.: Sequential responses in the prey-capture behavior of *Argiope argentata* (Fabricius). Web Building Spiders, AAAS Section on Zoology, New York, 1967 (in preparation, Amer. Zoologist 1968).

RUDALL, K. M.: Silk and other cocoon proteins. In: Comparative Biochemistry. Ed. M. FLORKIN and H. S. MASON, Acad. Press. N.Y., **4** b, 397—433 (1962).

SAINT RÉMY, G.: L'Étude du cerveau chez les arthropodes trachéates. Arch. Zool. exp. Gén. 2° Serie. T. V. bis. Suppl. **1**—**12**, 163—178, 265—270 (1887).

SALPETER, M., and C. WALCOTT: An electron microscopical study of a vibration receptor in the spider. Exp. Neurol. **2**, 232—250 (1960).

SAVORY, T. H.: The biology of spiders. London: Sidgwick and Jackson, Ltd. 1928.

— The spider's web. London: Frederick Warne & Co., Ltd. 1952.

— Instinctive living. London: Pergamon Press 1959.

— Spider webs. Sci. Amer. **202**, 114—125 (1960).

— Spiders, Men, and Scorpions. London: University Press 1961.

— Arachnida. London: Academic Press 1964.

SCHWARZ, R.: Versuche über die Beeinflussung des Nestbaues der Spinne *Zilla-x-notata* Cl. durch Stickoxydul und Äther. Arch. int. Pharmacodyn. **104**, 339—364 (1956).

SEIFTER, S., and P. M. GALLOP: The structure proteins. In: The Proteins, Vol. 4, p. 153—458. H. NEURATH (Ed.). New York: Academic Press 1966.

SEKIGUCHI, K.: On a new spinning gland found in geometric spiders and its functions. Ann. Zool. Jap. **25**, 394—399 (1952).

— Differences in the spinning organs between male and female spiders. Sci. Rep. Tokyo Kyoiku Daigaku, Sect. B. **8**, 23—32 (1955a).

— The spinning organs in sub-adult geometric spiders and their changes accompanying the last moulting. Sci. Rep. Tokyo Kyoiku Daigaku, Sect. B. **8**, 33—40 (1955b).

— Reduplication in spiders' eggs produced by centrifugation. Sci. Rep. Tokyo Kyoiku Daigaku, Sect. B. **8**, 187—280 (1957).

SIMON, M. E.: Études Arachnologiques. Memoirs **1**—**34** (1873).

SIMPSON, I.: Effect of some anesthetics on honey bees: nitrous oxide, carbon dioxide, ammonium nitrate, smoke fumes. Bee World **35**, 149—155 (1954).

SMITH, M. H.: The amino acid composition of proteins. J. theor. Biol. **13**, 261—282 (1966).

SPRONK, F.: Die Abhängigkeit der Netzbauzeiten der Radnetzspinne *Epeira diademata* und *Zilla-x-notata* von verschiedenen Außenbedingungen. Z. vergl. Physiol. Abt. C., **22**, 604—613 (1935).

SUZUKA, I., and K. SHIMURA: Biosynthesis of silk fibroin. 1. Incorporation of glycine-C^{14} into particles of posterior silk gland in vitro. J. Biochem. **47**, 551—554 (1960).

SZLEP, R.: On the plasticity of instinct of a garden spider (*Aranea diadema* L.), construction of a cobweb. Biol. exp. **16**, 5—22 (1952).

— Influence of external factors on some structural properties of the garden spider (*Aranea diademata*) web. Folia biol. (Praha) **6**, 287—299 (1958).

— Unusual web pattern made by an unfertilized female of the spider *Uloborus walckenaerius* Latreille. Nature (Lond.) **186**, 94 (1960).

— Developmental changes in the web spinning instinct of Uloboridae: Construction of the primary-type web. Behaviour **17**, 60—70 (1961a).

SZLEP, R.: Some differences in the running pattern in two spider families Uloboridae and Argiopidae. Bull. Res. Counc. Israel **9-B**, 1 (1961b).
— Change in the response of spiders to repeated web vibrations. Behaviour **23**, 203—239 (1964).
— The web-spinning process and web-structure of *Latrodectus tredecimguttatus*, *L. pallidus* and *L. revivensis*. Proc. Zool. Soc. (Lond.) **145**, 75—89 (1965).
TAMANO, N., and K. KURIAKI: Applicability of the silk worm for pharmacological studies acting on nervous systems, antimitotics and insecticides. Arch. int. Pharmacodyn. **132**, 49—59 (1961).
TILQUIN, A.: Sur l'orientation de l'argiope stationnant au centre de sa toile. J. Psychol. norm. path. **36**, 93—98 (1939).
— La Toile Géométrique des Araignées. Paris: Presses Universitaires de France 1942.
TUGENDHAT, B.: Feeding in conflict situations and following thwarting. Science **132**, 896—897 (1960).
— Hunger and sequential responses in the hunting behavior of salticid spiders. J. comp. physiol. Psychol. **58**, 167—173 (1964).
VAN DER KLOOT, W. G., and C. M. WILLIAMS: Cocoon construction by the cecropia silkworm. III. The alteration of spinning behavior by chemical and surgical techniques. Behaviour **6**, 233—255 (1954).
WALCOTT, C.: The effect of the web on vibration sensitivity in the spider *Achaearanea tepidariorum* (Koch). J. exp. Biol. **40**, 595—611 (1963).
— Vibration receptor of *Achaearanea tepidariorum*. Web-Building Spiders, AAAS Section on Zoology, New York, 1967 (in preparation, Amer. Zoologist, 1968).
—, and W. G. VAN DER KLOOT: The physiology of the spider vibration receptor. J. exp. Zool. **141**, 191—244 (1959).
WARBURTON, C.: The spinning apparatus of geometric spiders. Quart. J. micr. Sci. **31**, 29—39 (1890).
WARWICKER, J. O.: Comparative studies of fibroins. 2. The crystal structures of various fibroins. J. molec. Biol. **2**, 350—362 (1960).
— Comparative studies of fibroins. 5. X-ray examination of chemical resistant fractions of some silk fibroins. Biochim. biophys. Acta (Amst.) **52**, 319—328 (1961).
WELLS, F. L.: Behavior notes on *E. insularis* Hentz: Domestication, involution. J. genet. Psychol. **68**, 159 (1946).
WESTBERG, P.: Das Netz der Kreuzspinnen. Natur und Schule **4**, 27—33, 78—85, 116—129 (1905).
WESTRING, N.: Araneae Svecicae. Sumter et Litteris D. F. Gonnier 1861.
WIEHLE, H.: Beiträge zur Kenntnis des Radnetzbaues der *Epeiriden*, *Tetragnathiden* und *Uloboriden*. Z. Morpholog. u. Ökolog. der Tiere **8**, 468—537 (1927).
— In F. DAHL: Die Tierwelt Deutschlands, 44. Teil: Spinnentiere oder Arachnoidea. Jena: Fischer 1956.
WILDER, B. G.: The Nets of *Epeira*, *Nephila* and *Hyptiotes* (Mithras). Amer. Ass. Advance. Sci. Proc. Salem 264—274 (1874).
WILLIAMS, G.: Seasonal and diurnal activity of harvestmen and spiders in contrasted habitats. J. Anim. Ecol. **31**, 23—42 (1962).
WILSON, R. S.: The structure of the dragline control valves in the garden spider. Quart. J. micr. Sci. **103**, 549—555 (1962a).
— The control of the dragline spinning in the garden spider. Quart. J. micr. Sci. **103**, 557—571 (1962b).
— The pedicel of the spider *Heteropoda venatoria*. J. Zool. **147**, 38—45 (1965).
— The control of dragline spinning in certain spiders. Web-Building Spiders, AAAS Section on Zoology, New York, 1967 (in preparation, Amer. Zoologist, 1968).
WITT, P. N.: Verschiedene Wirkung von Coffein und Pervitin auf den Netzbau der Spinne. Helv. physiol. Acta **7**, C 65 (1949).
— d-Lysergsäure-diaethylamid (LSD 25) im Spinnentest. Experientia (Basel) **7**, 310 (1951).
— Ein einfaches Prinzip zur Deutung einiger Proportionen im Spinnennetz. Behaviour **4**, 172—189 (1952).
— Ein biologischer Nachweis von Adrenochrom und seine mögliche Anwendung. Helv. physiol. pharmacol. Acta **12**, 327—337 (1954).
— Die Wirkung einer einmaligen Gabe von Largactil auf den Netzbau der Spinne *Zilla-x-notata*. Mschr. Psychiat. Neurol. **129**, 123—128 (1955).
— Die Wirkung von Substanzen auf den Netzbau der Spinne als biologischer Test. Berlin-Göttingen-Heidelberg: Springer 1956a.

Witt, P. N.: Der Netzbau der Spinne als Test zur Prüfung zentralnervös angreifender Substanzen. Arzneimittel-Forsch. **6**, 628—635 (1956 b).
— Vergleich der Wirkung von Xylopropamin und Pervitin auf den Netzbau der Spinne. Arch. int. Pharmacodyn. **180**, 143—152 (1956 c).
— The identification of small quantities of hallucinatory substances in body fluids with the spider test. Psychopathology, a source book, ed. Reed et al. Cambridge: Harvard University Press 1958.
— Effects of psilocybin on web-building behavior of spiders. Pharmacologist **2**, 2 (1960).
— Web building. The Encyclopedia of the Biological Sciences, ed. P. Gray, p. 1072—1073. New York: Reinhold Publ. Co. 1961.
— Effects of atropine on spiders' web building behavior and thread production. Fed. Proc. **21**, 80 (1962).
— Environment in relation to the behavior of spiders. Arch. environm. Hlth **7**, 4—12 (1963).
— Do we live in the best of all worlds? Spider webs suggest an answer. Perspect. Biol. Med. **8**, 475—487 (1965).
— Behavioral consequences of CNS lesions in *Araneus diadematus* Cl., Web-Building Spiders, AAAS Section on Zoology, New York, 1967 (in preparation, Amer. Zoologist, 1968).
—, and R. Weber: Biologische Prüfung des Urins von drei Kranken mit akut psychotischen Zustandsbildern auf pathogene Substanzen mit dem Spinnentest. Mschr. Psychiat. Neurol. **132**, 193—207 (1956).
—, and R. Baum: Changes in orb webs of spiders during growth (*Araneus diadematus* Clerck and *Neoscona vertebrata* McCook). Behaviour **16**, 309—318 (1960).
—, and C. F. Reed: Spider-web building. Measurement of web geometry identifies components in a complex invertebrate behavior pattern. Science **149**, 1190—1197 (1965).
— —, and F. K. Tittel: Laser lesions and spider web construction. Nature (Lond.) **201**, 150—152 (1964).
Wolff, D., u. U. Hempel: Versuche über die Beeinflussung des Netzbaues von *Zilla-x-notata* durch Pervitin, Scopolamin und Strychnin. Z. vergl. Physiol. **33**, 497—528 (1951).

Author Index

Subject Index

Druck von J. P. Peter, Gebr. Holstein, Rothenburg o. d. Tbr.

M